CW00794607

Lea Stein® Jewelry

Judith Just

4880 Lower Valley Road, Atglen, PA 19310 USA

All designs and trade mark Lea Stein Paris® are protected. For further requirement, please contact leastein@club-internet.fr

This book was derived from author's independent research and was not authorized by Lea Stein.

All content information, including valuations, has been compiled from the most reliable resources, and every effort has been made to eliminate errors and questionable data. Nevertheless, the possibility of error in a work of this scope always exists. Persons who feel they have discovered errors are invited to write and inform us so they may be corrected in subsequent editions.

The current values in this book should be used only as a guide. They are not intended to set prices which vary from one section of the country to another, and are affected by condition as well as demand. Neither the author nor the publisher assumes responsibility for any losses that might be incurred as a result of consulting this guide.

Cover: Photograph of butterflies *courtesy of Nathalie Bernhard.*
Title page, right: Photograph *courtesy of Lea Stein.*

Copyright © 2001 by Judith Just
Library of Congress Control Number: 2001090452

All rights reserved. No part of this work may be reproduced or used in any form or by any means—graphic, electronic, or mechanical, including photocopying or information storage and retrieval systems—without written permission from the copyright holder.

"Schiffer," "Schiffer Publishing Ltd. & Design," and the "Design of pen and inkwell" are registered trademarks of Schiffer Publishing Ltd.

Designed by Bonnie M. Hensley
Cover design by Bruce M. Waters
Type set in Nuptial BT/Souvenir Lt BT

ISBN: 0-7643-1381-9
Printed in China
1 2 3 4

Published by Schiffer Publishing Ltd.
4880 Lower Valley Road
Atglen, PA 19310
Phone: (610) 593-1777; Fax: (610) 593-2002
E-mail: Schifferbk@aol.com
Please visit our web site catalog at **www.schifferbooks.com**

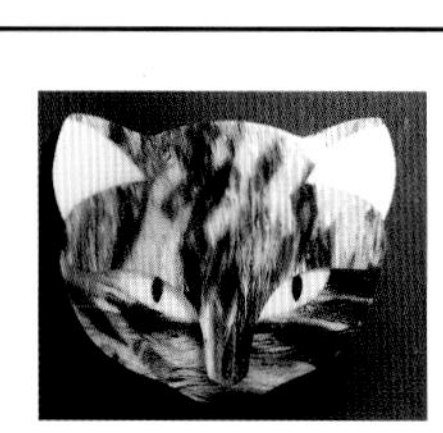

In Europe, Schiffer books are distributed by Bushwood Books
6 Marksbury Avenue Kew Gardens
Surrey TW9 4JF England
Phone: 44 (0) 20-8392-8585; Fax: 44 (0) 20-8392-9876
E-mail: Bushwd@aol.com
Free postage in the UK. Europe: air mail at cost.

This book may be purchased from the publisher. Include $3.95 for shipping. Please try your bookstore first. We are always looking for people to write books on new and related subjects. If you have an idea for a book please contact us at the Atglen, PA address. You may write for a free catalog.

Dedication

To my husband, Hal:
my perfect love, my dearest friend, my inspiration.

Acknowledgements

This book is only the result of these special people who were most generous with their support, the sharing of their collections, energies, expertise, talent, and friendship. It is my pleasure to acknowledge them individually, and to gratefully extend my thanks and appreciation to Lea Stein, and Fernand Steinberger, Nathalie Bernhard (nbernhar@noos.fr), Anita L. Bielecki, Gail Crockett (@aunttink.com), Missy DeBellis, Adele Gordon, Linda Heck, Ethel Malino, Susan Marks of OhMy!!, Renee Piche, Remembrances of Things Past (376 Commercial Street, Provincetown, Massachusetts), Dale Rhoads, and Nancy Schiffer.

Photography Credits

Photography of the Lea Stein Collection courtesy of Lea Stein. Photography of the Nathalie Bernhard Collection courtesy of Nathalie Bernhard. Photography of the "Private Collector" collection by Anita L. Bielecki. All other photography by Judith Just.

Contents

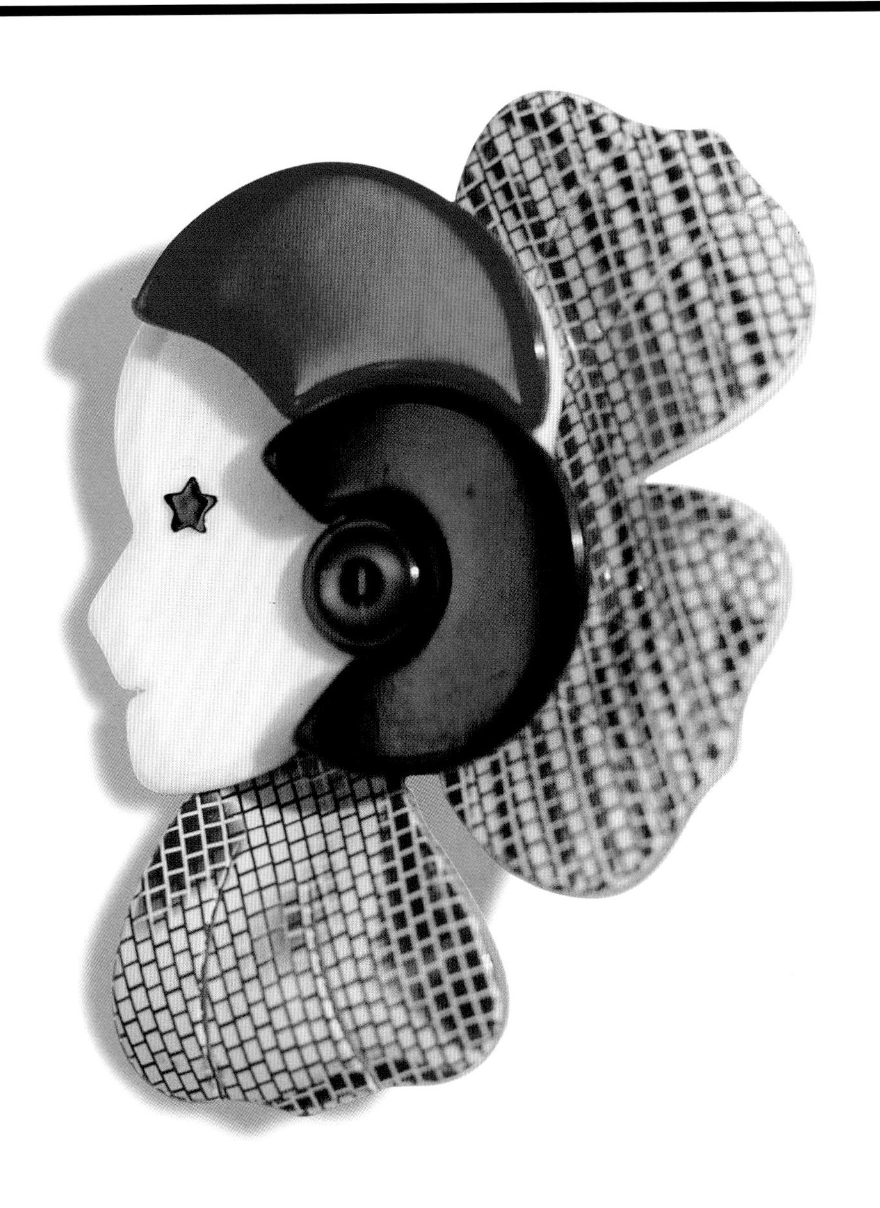

Introduction

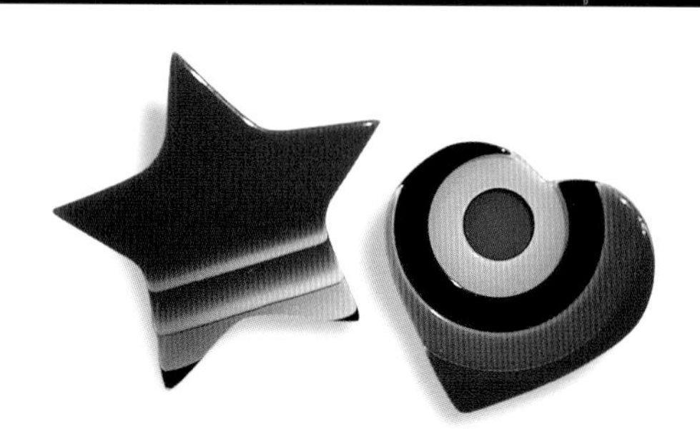

The urge to embellish and the love of ornamental effect are basic to human nature. People of every society have tried to invent beautiful objects to give meaning and importance to an often drab reality. Ornament is essentially free. Free to move the eyes, to intrigue the mind, to rest the soul; free to simply delight us, as do the creations of Lea Stein.

Lea Stein was born in Paris. She married Fernand Steinberger in the 1950s. She was the designer and Fernand invented the laminated celluloid process they use to produce the remarkable, many layered, multi-colored jewelry and the accessories that appear in this book. Together, they created jewelry that has been appreciated for its individuality and innovation throughout the world.

The technical formula for Lea Stein jewelry must be credited to the genius of her husband, Fernand Steinberger. Manufacturing the designs is extremely complex and time consuming work. One of the important components is a well kept secret. Many laminated cellulose acetate sheets are layered with pieces of lace or imaginative fabrics with iridescent textures and patterns. The combinations are then blended and baked. Lea Stein uses this layered or stacked method to create her unusual sculptural designs.

The architecture and construction of this whimsical world is in a class by itself. Her concept of color, form, and design are powerful and unique. Lea Stein has shown through her extraordinary talent that plastic jewelry can be elevated to the highest levels of art. Through originality she has achieved lasting prominence in a highly competitive field. Now let the fun begin!

Chapter 1

Collecting Lea Stein Paris® Jewelry

Today, collecting Lea Stein's jewelry is truly an addiction. Since the 1960s, Lea Stein has designed and produced an extensive collection of pins, necklaces, earrings, bracelets, ornamental combs, rings, picture frames, jewelry boxes, buttons, and accessories. The majority of her designs consist of a variety of whimsical, colorful pins: children, men, women, animals, birds, flowers, boats, hats, geometric designs, and other subjects, which are streamlined in the Art Deco style. The buttons, accessories, and serigraphy pins are rather rare and difficult to find today.

One should always be aware that, as with any item of collectible significance, there is always the possibility of imitations. To be certain of acquiring authentic Lea Stein pieces, study the items in this book carefully and the item itself before you purchase it. Be certain that the seller has an impeccable reputation and is willing to accept a return, should you discover it is not authentic, because sometimes, although rarely, even a seller may be fooled.

Marks

Some of the early pins from the 1960s are unsigned and a v-shaped clasp was used. Later, and as it is today, the v-shaped clasp was signed "Lea Stein Paris." The clasp is attached to the back of the pin by a method of heat, mounting the center with melted plastic. Some of the later versions are riveted.

Early bracelets, earrings, necklaces, rings and other items do not always bear her signature, but can be recognized by her unmistakable style. Some of the bracelets designed in early years are being signed today, before they are sold, with her engraved signature.

Values

The current retail values should be used only as a guide. Be aware that prices vary from one place to another, and from one dealer to another. A low and a high price spread has been worked out for the benefit of the reader. If no value is stated, it is simply because no price is available. Be aware that prices differ widely when comparing domestic and overseas markets. All of the values here were determined by exhibitors, collectors, and sellers across the United States.

Laminated and layered bracelets, c. 1970s. *Courtesy of OhMy!! and Author.* $325-700.

Ornamental Hair Combs, c. 1968-1980. *Lea Stein Collection, Paris.* $100-175.

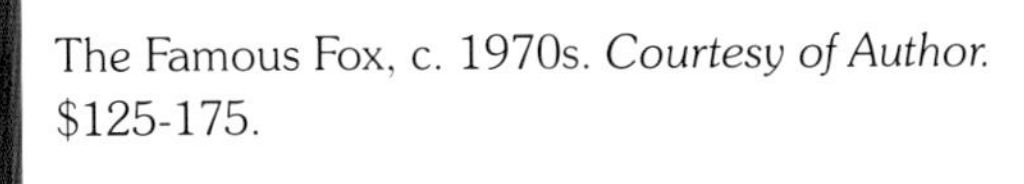

The Famous Fox, c. 1970s. *Courtesy of Author.*
$125-175.

Three dog pins

Chapter 2

Buttons

Lea Stein's earliest work in the 1960s included unusual buttons that were bought by specialty stores in France and other countries in Europe. They were sought after by many French fashion designers.

Button collections from the 1960s. *Courtesy of Nathalie Bernhard.*

Chapter 3

Early Serigraphy Pins

Serigraphy pins were also among Lea Steins' early designs. The pins sometimes depicted simplified Art Deco designs, floral designs, and exquisite women wearing stylish hats and clothing from the 1920s, or earlier days. The pins are works of art and look like fine, miniature paintings.

Three serigraphy pins of women's faces, c. 1960s. *Courtesy of Nathalie Bernhard.* $200-300 each.

Serigraphy pins of triangle with floral design, Art Deco design, and young girl with hat and flowers, c. 1960s. *Courtesy of Lea Stein collection, Paris.* $200-300 each.

Serigraphy pin of woman with flowers in faux tortoise frame, c. 1960s. *Courtesy of private collector. Photo by Anita L. Bielecki.* $200-300.

Two serigraphy pins, one of woman in faux tortoise frame and little girl in pearlized swirl frame, c. 1960s. *Courtesy of Lea Stein Collection, Paris.* $200-300 each.

Two serigraphy pins, c. 1960s. At left is Julius Caesar with "Veni Vidi Vici" printed on it. *Courtesy of Nathalie Bernhard.* $200-300 each.

Three serigraphy pins of women framed in faux tortoise, c. 1960s. *Courtesy of Nathalie Bernhard.* $200-$350 each.

Two serigraphy pins of multi-colored flowers framed in faux tortoise, c. 1960s. *Courtesy of Nathalie Bernhard.* $250-$350 each.

Chapter 4

"Isle of Children"

In the 1970s, Lea Stein bought the rights to design and produce pins of all the characters in the popular French TV series for children, L'ILE AUX ENFANTS ("Isle of Children"). Today, these pins are extremely collectible.

2 Gribouille Happpy Dog pins. This and all that follow in this chapter are from the French televison series 'Isle of Children,' c. 1975. 2" x 1.25". *Courtesy of OhMy!!* $125-175 each.

Pair of Tifins, mascot of the television series, c. 1975. 1 3/4". *Courtesy of Nathalie Bernhard.* $125-175 each.

Casimir a character pin, c. 1975. 2". *Courtesy of OhMy!!* $125-175.

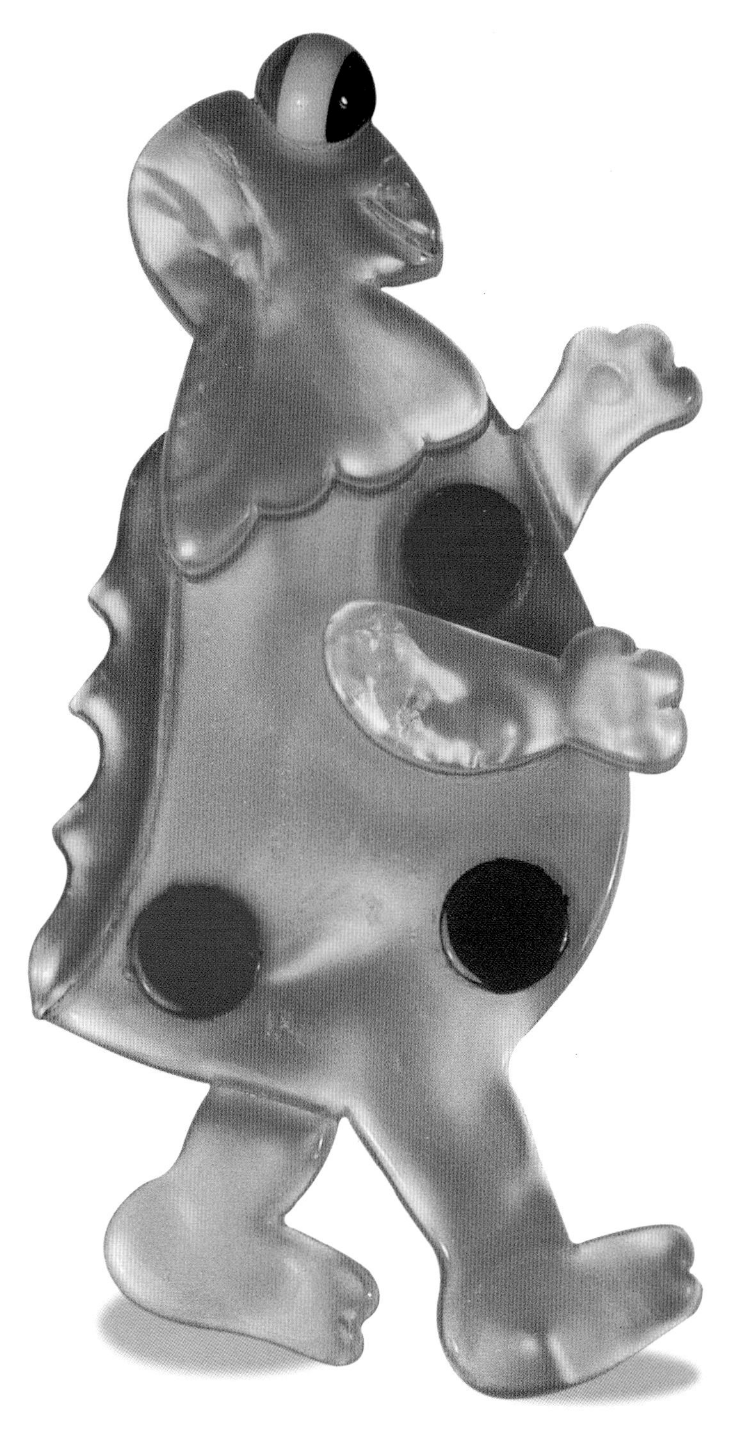

Pair of Maya the Bee pins, c. 1975. 1 3/4". *Courtesy of OhMy!!* $125-175 each.

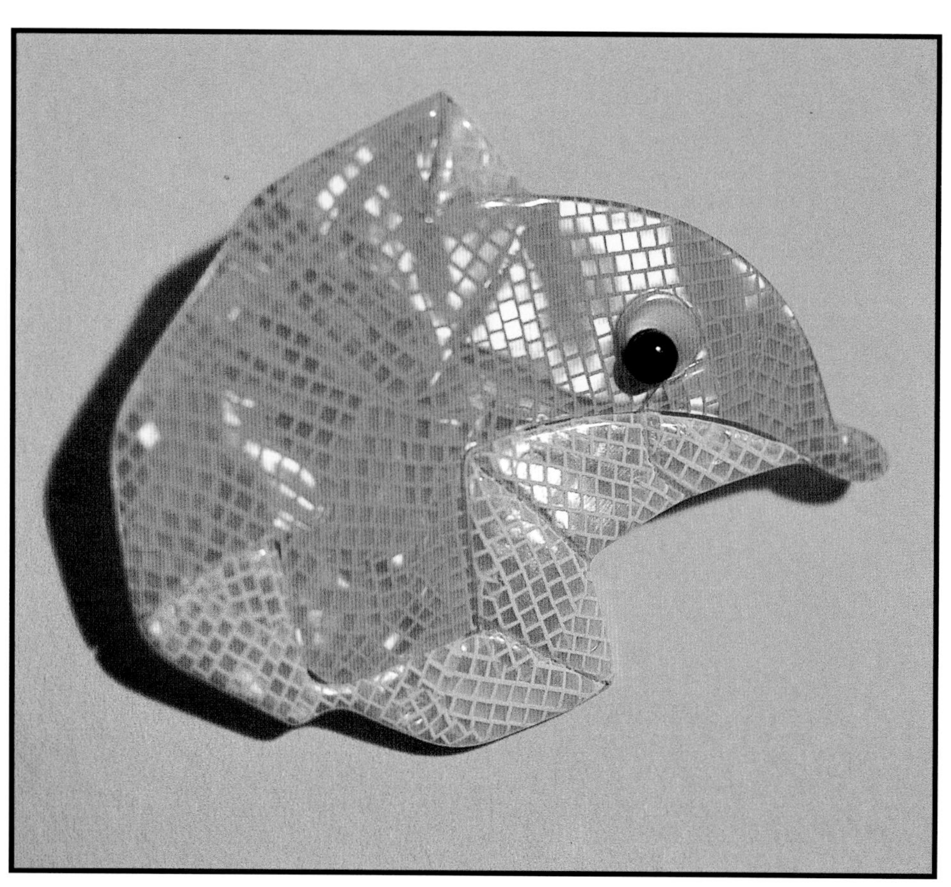

Character pin, c. 1975. 2". *Courtesy of OhMy!!* $125-175.

Léonard the frog pin, c. 1975. 2 1/8". *Courtesy of OhMy!!* $125-175.

Calimero the baby bird with egg shell on head pin, c. 1975. 2". *Courtesy of private collector. Photo by Anita L. Bielecki.* $125-175.

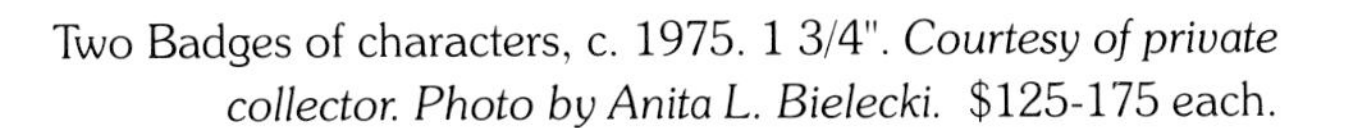

Two Badges of characters, c. 1975. 1 3/4". *Courtesy of private collector. Photo by Anita L. Bielecki.* $125-175 each.

Chapter 5

People Pins

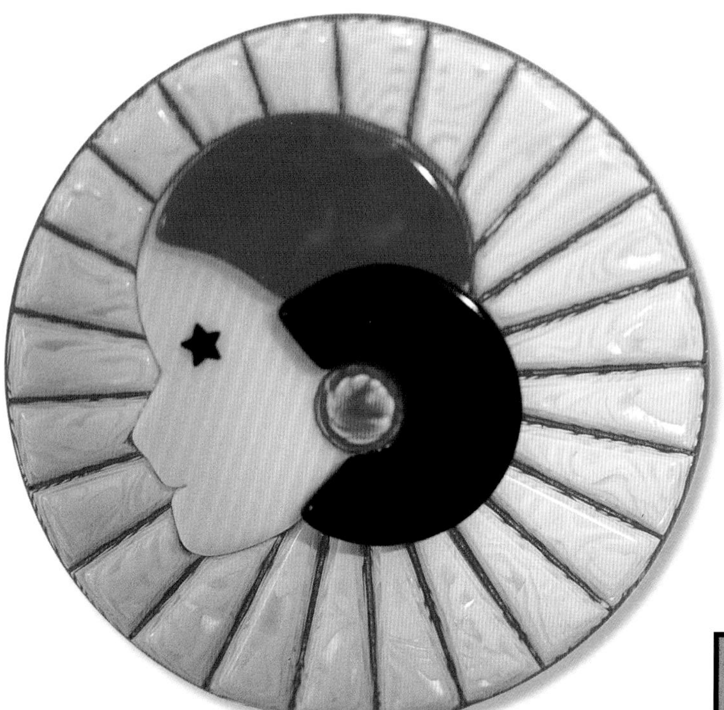

Moderne woman pins in multi-colors, also known as Colerette, c. 1968-1980. 2 1/4". *Courtesy of OhMy!! and author.* $125-175.

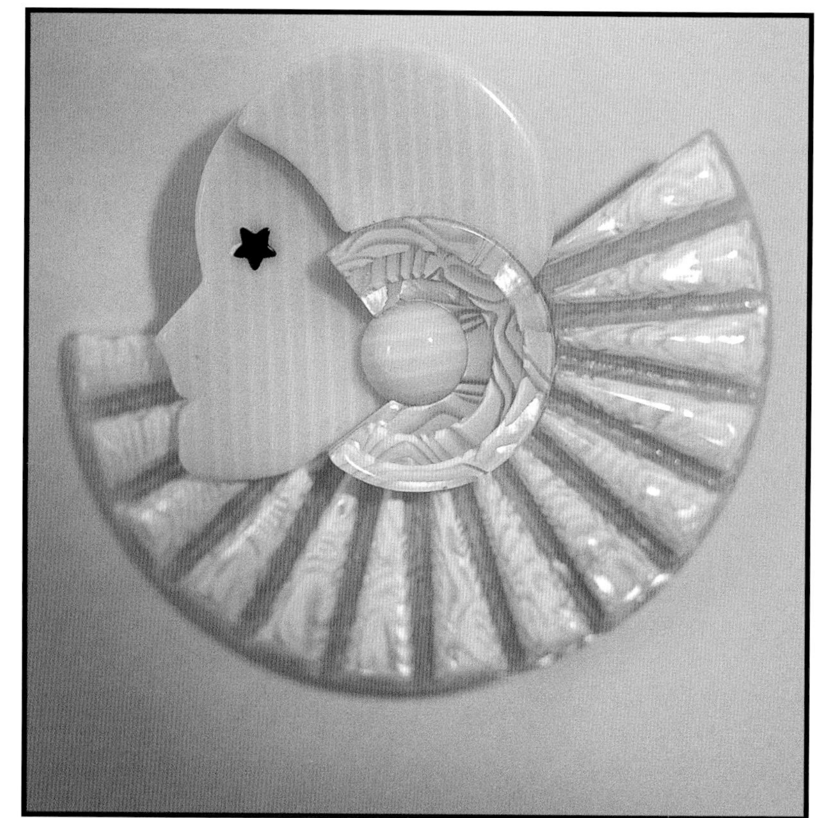

Pair of flapper pins known as Colombine, c. 1968-1980. 2 1/8". *Courtesy of OhMy!!* $125-175 each.

Pair of flapper pins known as Sauvage, c. 1968-1980. 1 7/8". *Courtesy of Remembrances of Things Past, Provincetown, MA.* $125-175 each.

Flapper pins known as Corolle, c. 1968-1980. 2 3/8". *Courtesy of OhMy!!* $125-175.

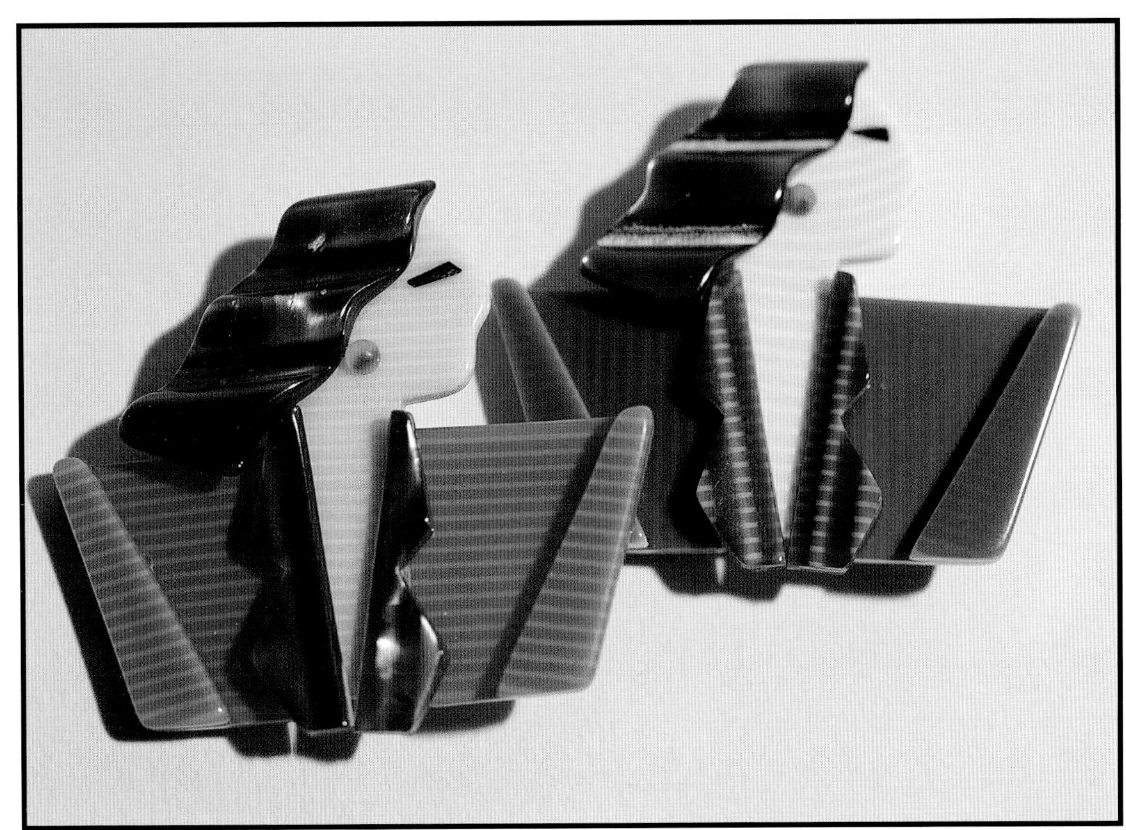

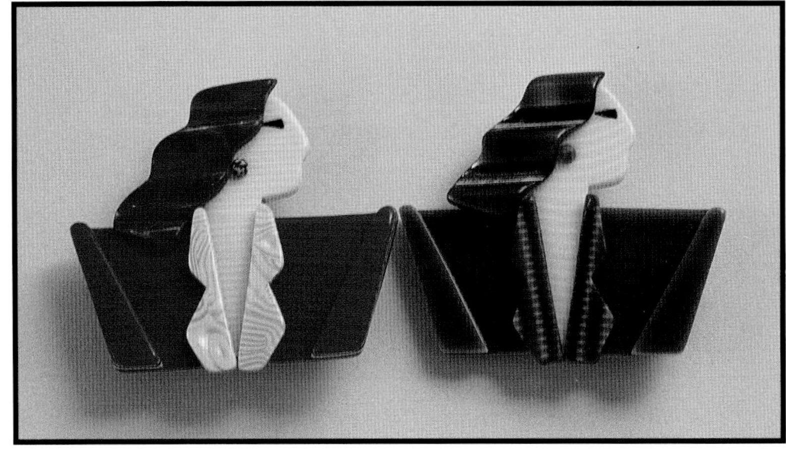

Pair of Joan Crawford pins, also known as Carmen, c. 1968-1980. 2". *Courtesy of Dale Rhoads and author.* $125-175 each.

Geometric pin with rhinestones and woman with saxaphone pin, c. 1968-1980. *Courtesy of Nathalie Bernhard.* $125-175 each.

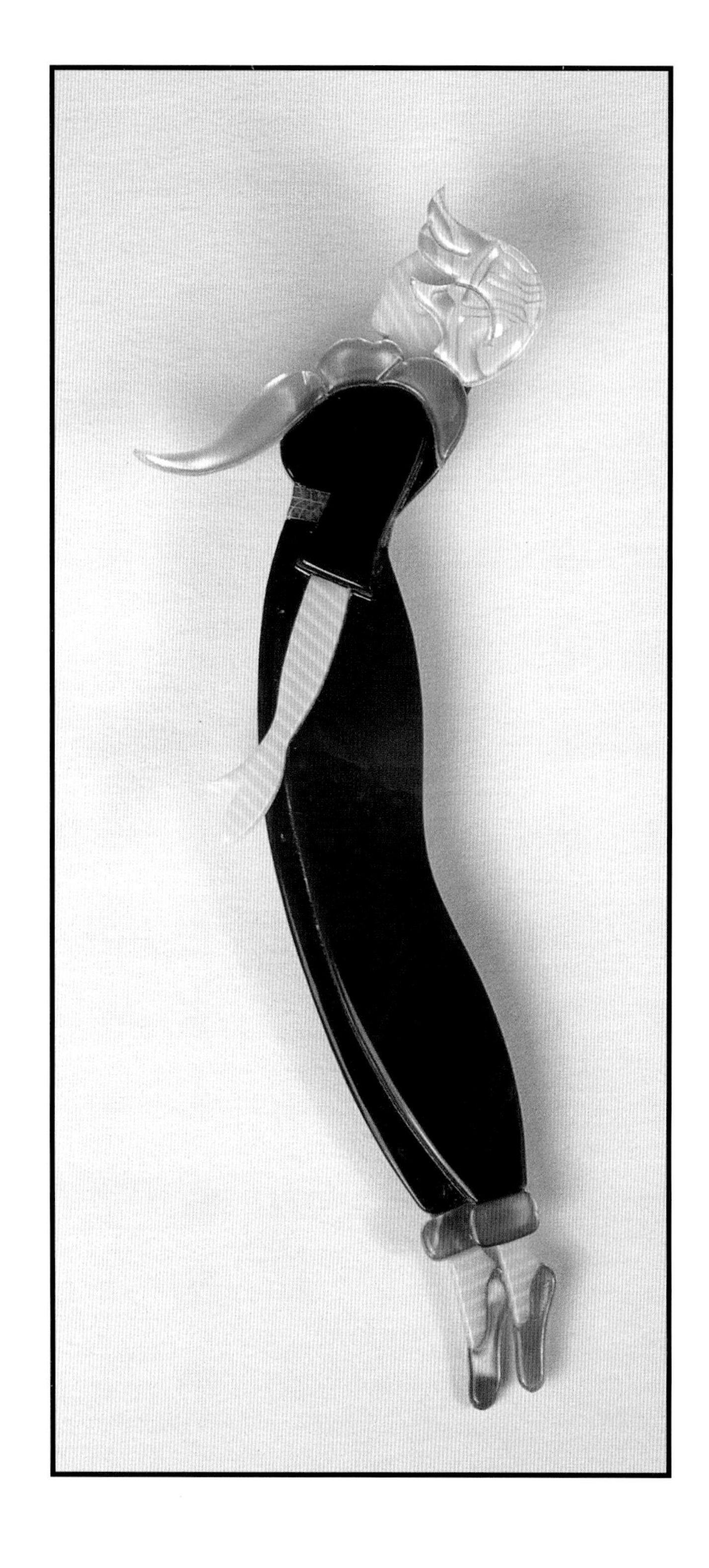

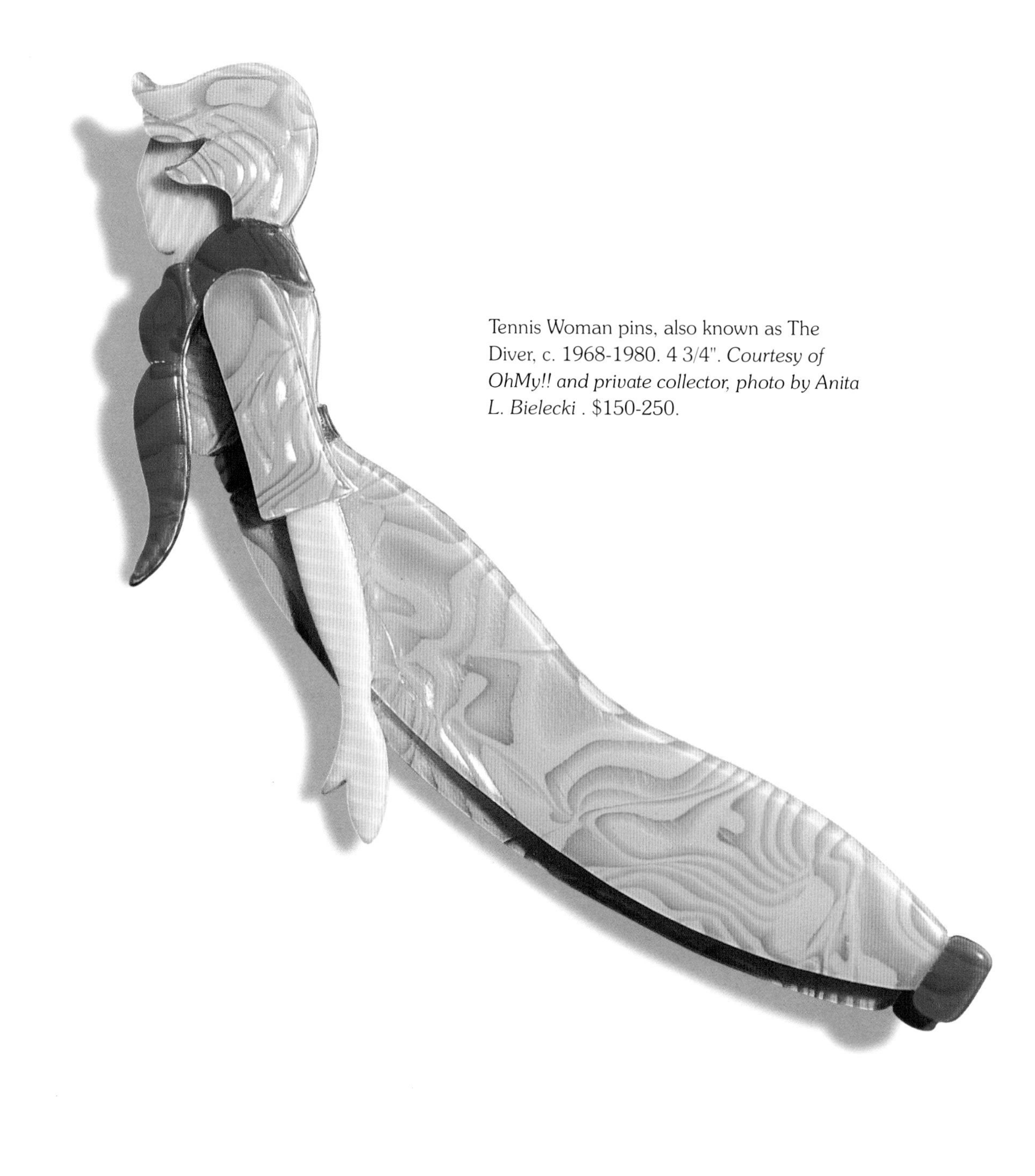

Tennis Woman pins, also known as The Diver, c. 1968-1980. 4 3/4". *Courtesy of OhMy!! and private collector, photo by Anita L. Bielecki* . $150-250.

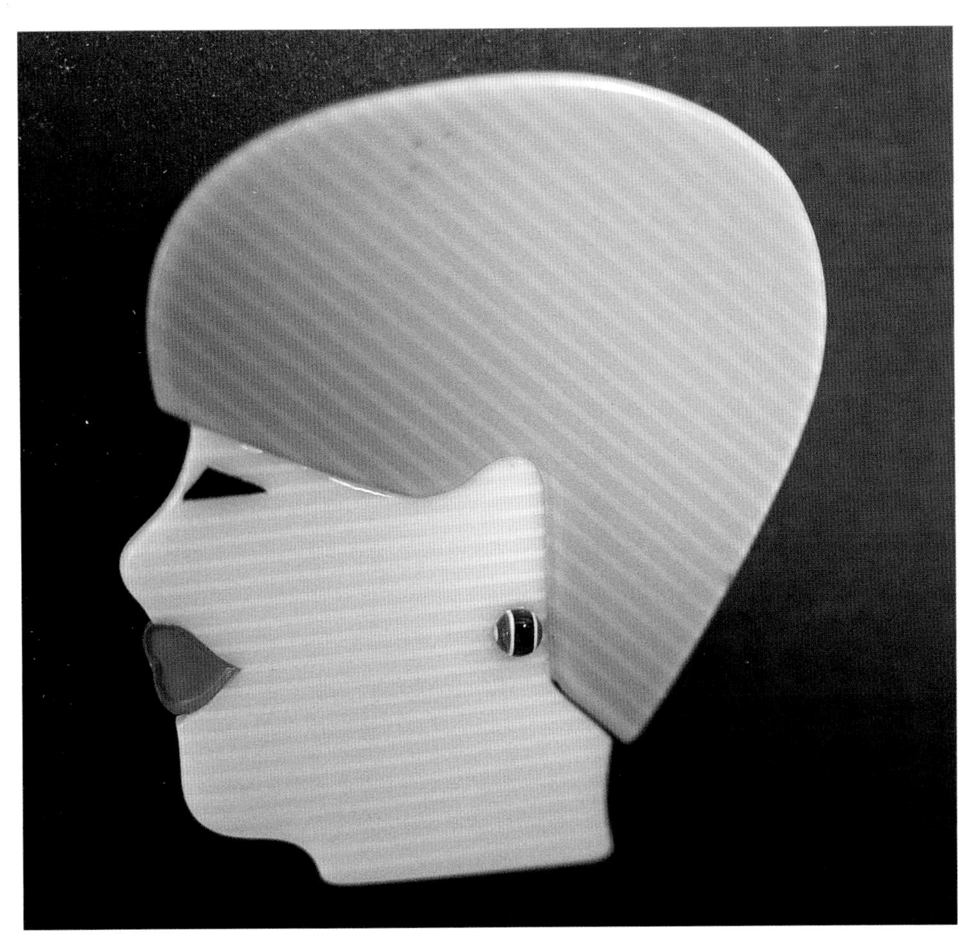

Deco lady pin, c. 1968-1980. 2".
Courtesy of author. $125-175.

Léontine, the young girl pin, c. 1968-1980.
1 3/4". *Courtesy of private collector. Photo by Anita L. Bielecki.* $125-175.

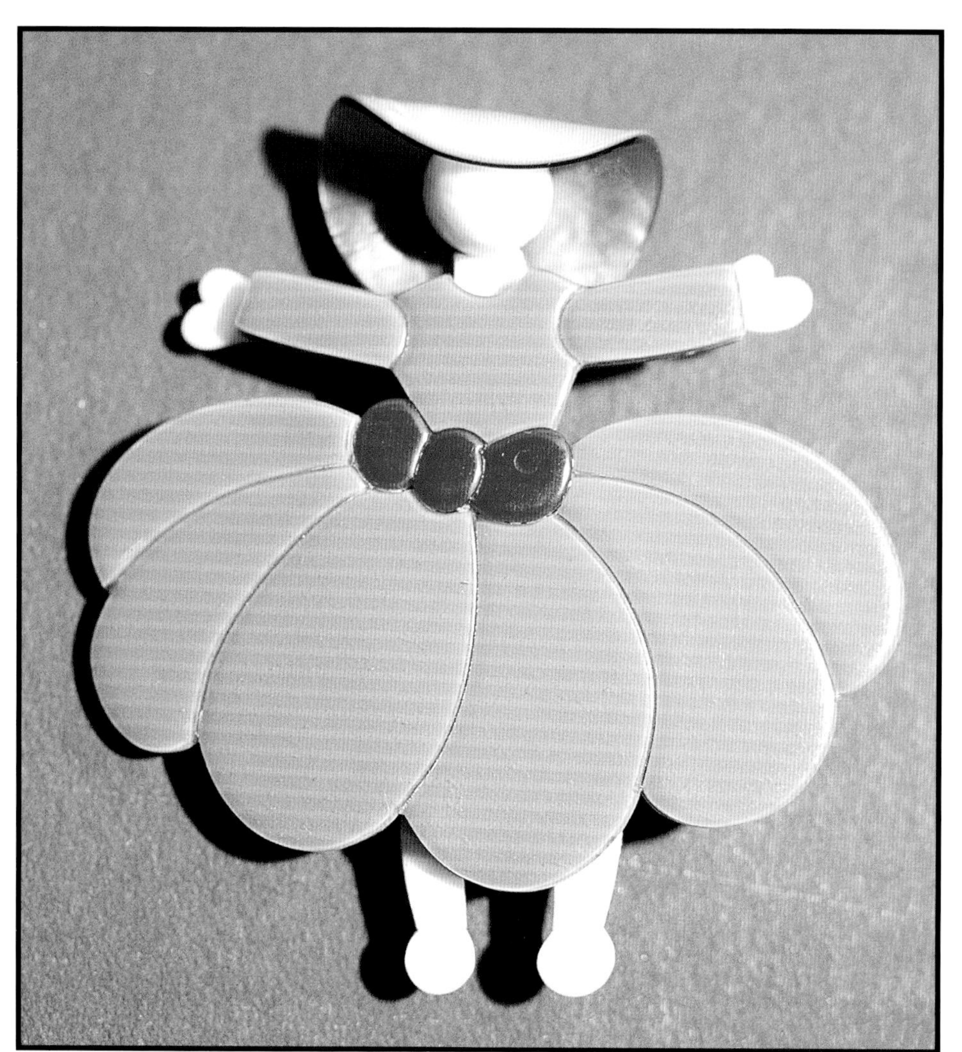

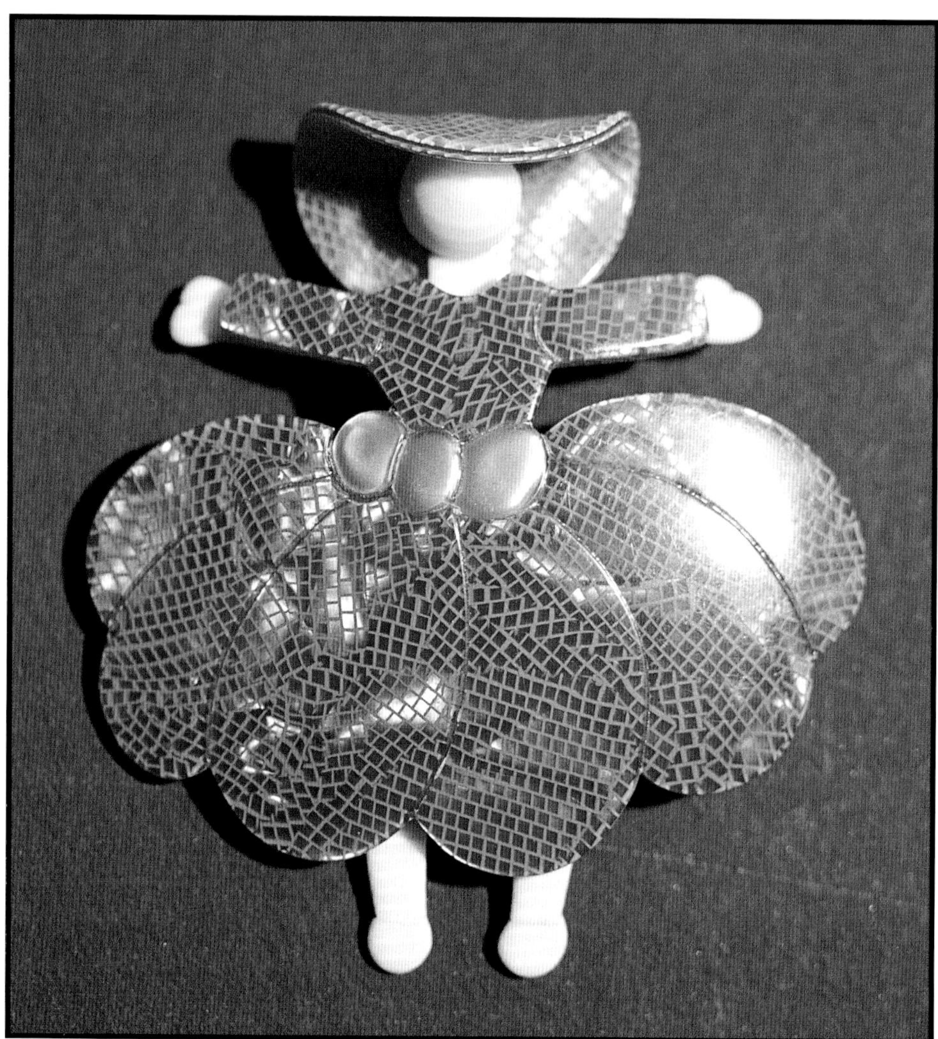

Scarlet O'Hara pins in olive and red, and in purple, also known as the Ballerina, c. 1968-1980. 2 1/2". *Courtesy of OhMy!!* $125-175.

Little girl with hat pin and little girl with ribbon pin, c. 1968-1980. 1 7/8". *Courtesy of OhMy!!* $125-175 each.

Pair of Scarlet O'Hara pins, one in grey and the other in coral, also known as the Ballerina, c. 1968-1980. 2 1/2". *Courtesy of private collector. Photo by Anita L. Bielecki. $125-175.*

Scarlet O'Hara pin in ivory, also known as the Ballerina, c. 1968-1980. 2 1/2". *Courtesy of OhMy!!* $125-175.

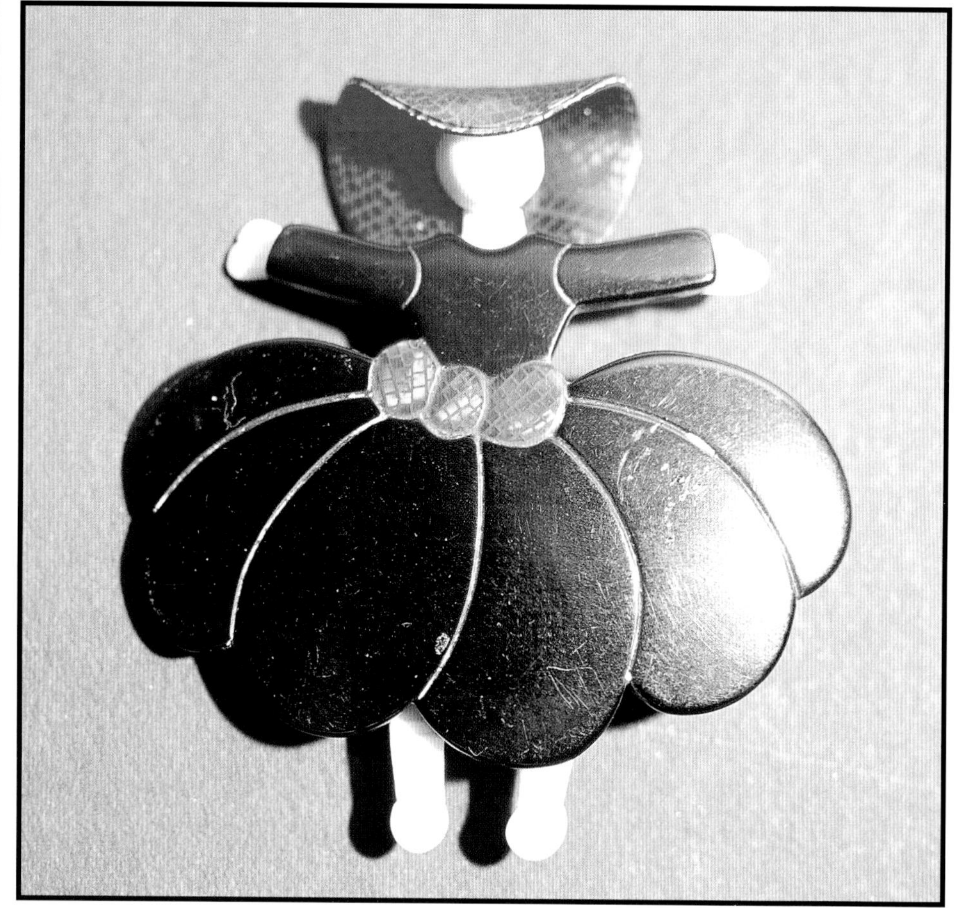

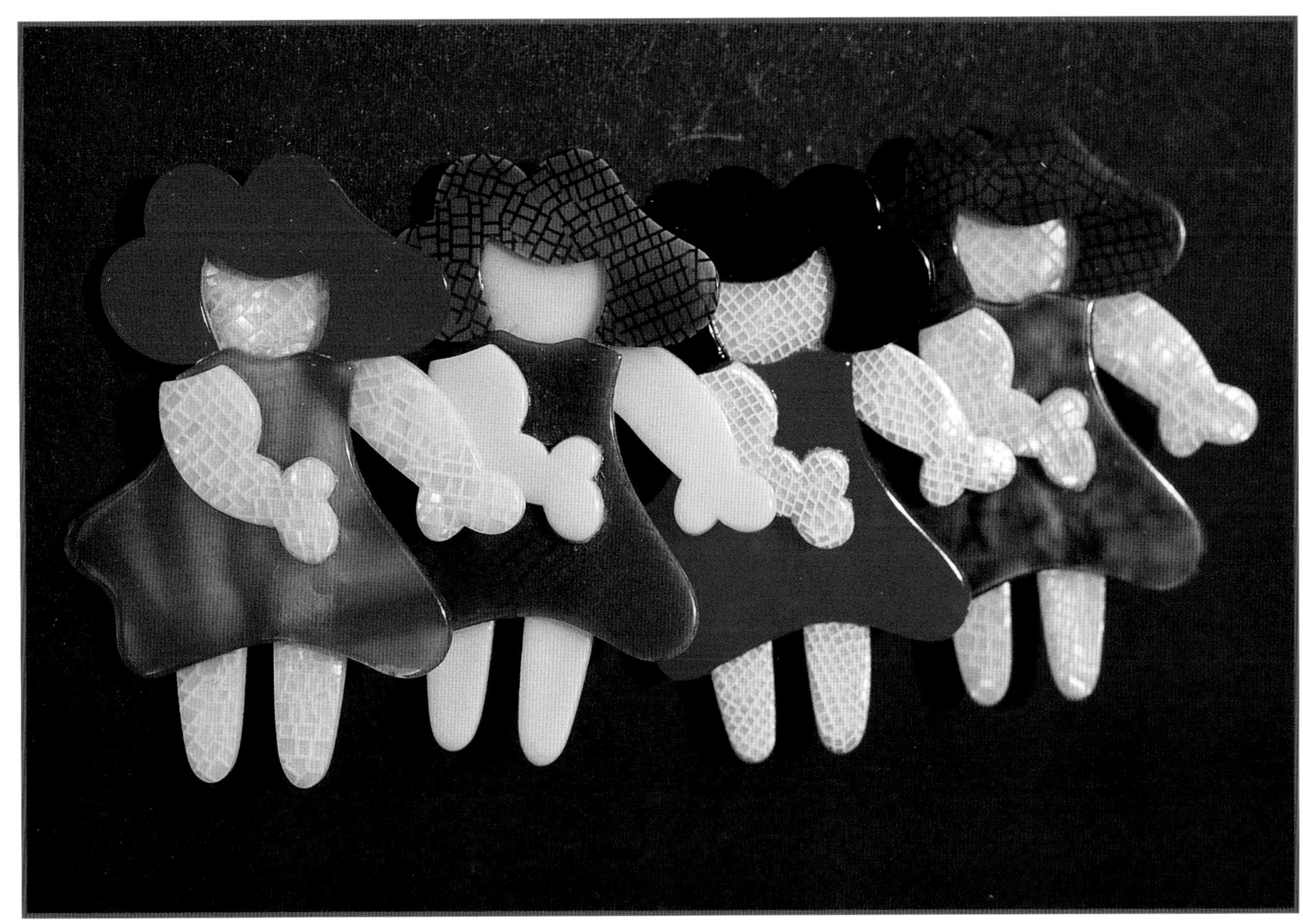

Four little dancing girl pins in multi-colored dresses, c. 1968-1980. 2 1/4". *Courtesy of OhMy!!* $125-175 each.

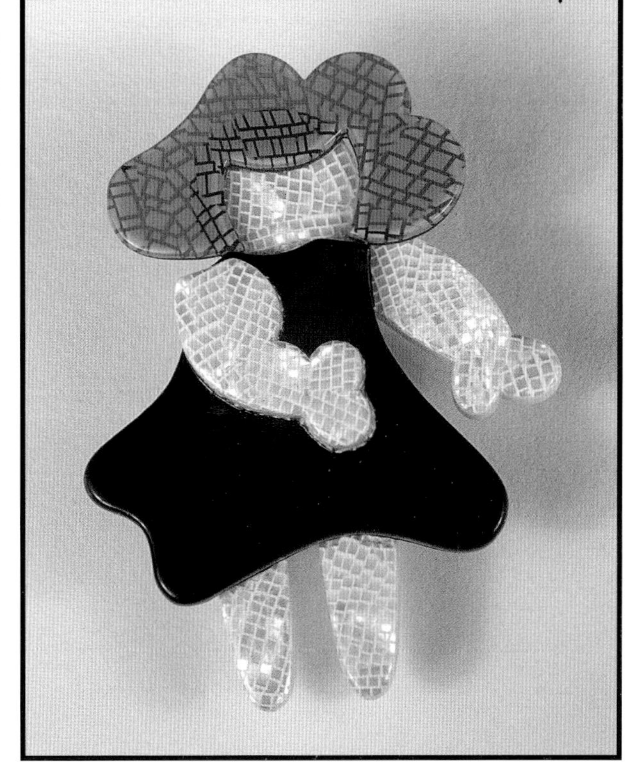

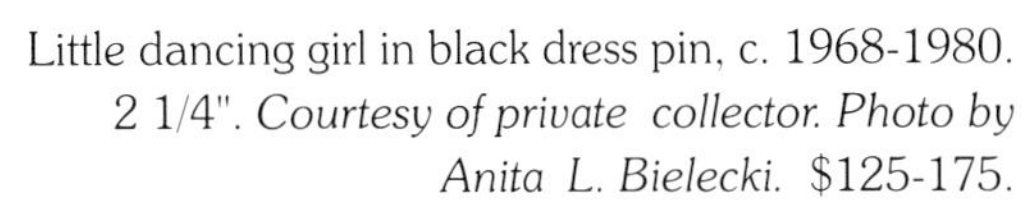

Little dancing girl in black dress pin, c. 1968-1980. 2 1/4". *Courtesy of private collector. Photo by Anita L. Bielecki.* $125-175.

Trio of little girl pins: girl on left in glitter dress and hat, little girl in middle sometimes referred to as Orphan Annie, dancing girl on right wears tie-dyed dress, c. 1968-1980. *Courtesy of author.* $125-175 each.

Rainbow and umbrella pins, c. 1968-1980. 1 1/2". *Courtesy of Dale Rhoads; and private collector, photo by* Anita L. Bielecki. $125-175 each.

Two trio of umbrellas pins, c. 1968-1980. 2". *Courtesy of OhMy!!* $125-175 each.

Group of sailor pins, c. 1968-1980. 2 3/8". *Courtesy of OhMy!! and author.* $125-175 each.

Brown and tan sailor pin, c. 1968-1980. *Courtesy of private collector. Photo by Anita L. Bielecki.* $125-175.

Rhinestone anchor pin and rhinestone sailor pin, c. 1968-1980. *Courtesy of Nathalie Bernhard.* $125-175 each.

French sailor pin, c. 1968-1980. 2 1/2". *Courtesy of OhMy!!* $125-175.

Chauffeur pin and stickpin, c. 1968-1980. Head only 1 5/8". *Courtesy of OhMy!!* $125-175 each.

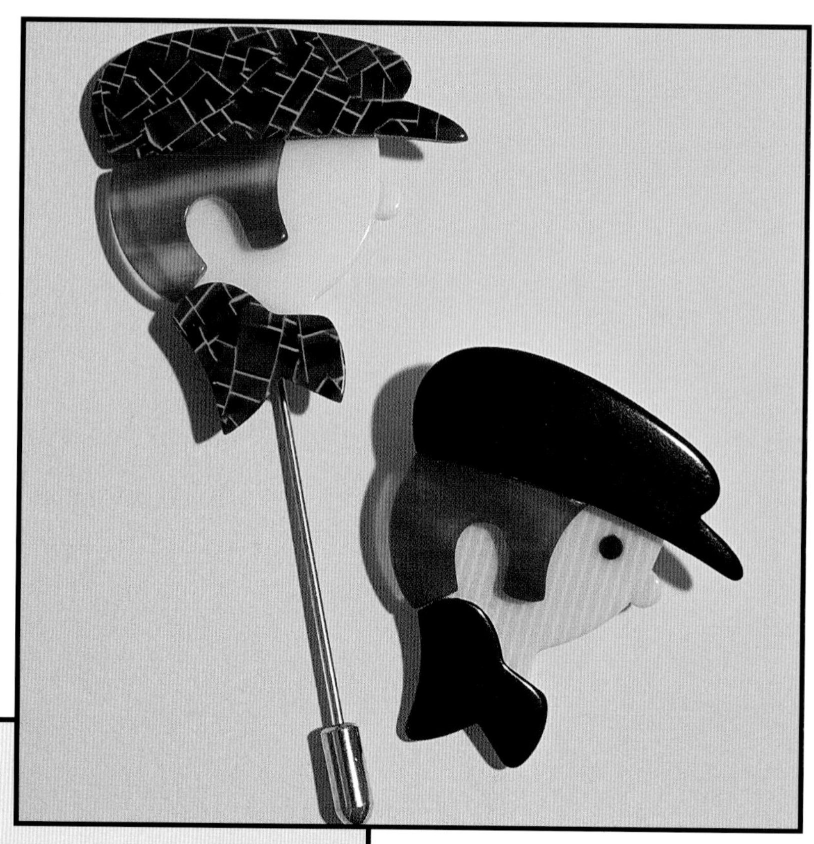

Group of French sailor pins, c. 1968-1980. 2 3/8". *Courtesy of Nathalie Bernhard.* $125-175 each.

The Benjamin pin, c. 1968-1980. 2 3/4"
Courtesy of private collector. Photo by Anita L. Bielecki. $125-175.

The Farmer pin, c. 1968-1980. 2 3/8".
Courtesy of Dale Rhoads. $125-175.

Pair of skateboarder pins, one with rhinestones, c. 1968-1980. 1 3/4" *Courtesy of OhMy!! and Dale Rhoads.* $125-175 each.

Pair of Elvis Presley pins, c. 1968-1980. 2 3/4". *Courtesy of OhMy!!* $125-200.

Two rare Chinese men pins, c. 1968-1980. 2 3/4" and 2". *Courtesy of Nathalie Bernhard and Dale Rhoads.* $250-375 each.

Red and silver Indian head pin, c. 1968-1980. 2 1/8". *Courtesy of author.* $125-175.

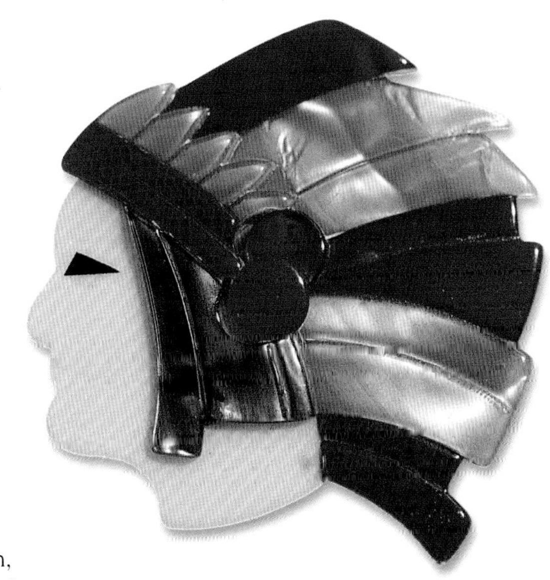

Purple and silver Indian head pin, c. 1968-1980. 2 1/8". *Courtesy of OhMy!!* $125-175.

Chapter 6

The Famous Foxes

The most famous of Lea Stein's designs to date is the 3-D fox pin, with slanted eyes and looped tail. This pin was designed in the 1960s and is still being produced in a wide range of colors and patterns. The back of this pin is usually a solid color in later versions. This is a most remarkable design and, whether an earlier or later version, an important collectible piece. Other Lea Stein designs are usually limited in production.

Five famous fox pins looking one another over, c. 1960s to present. 4" $125-175 each.

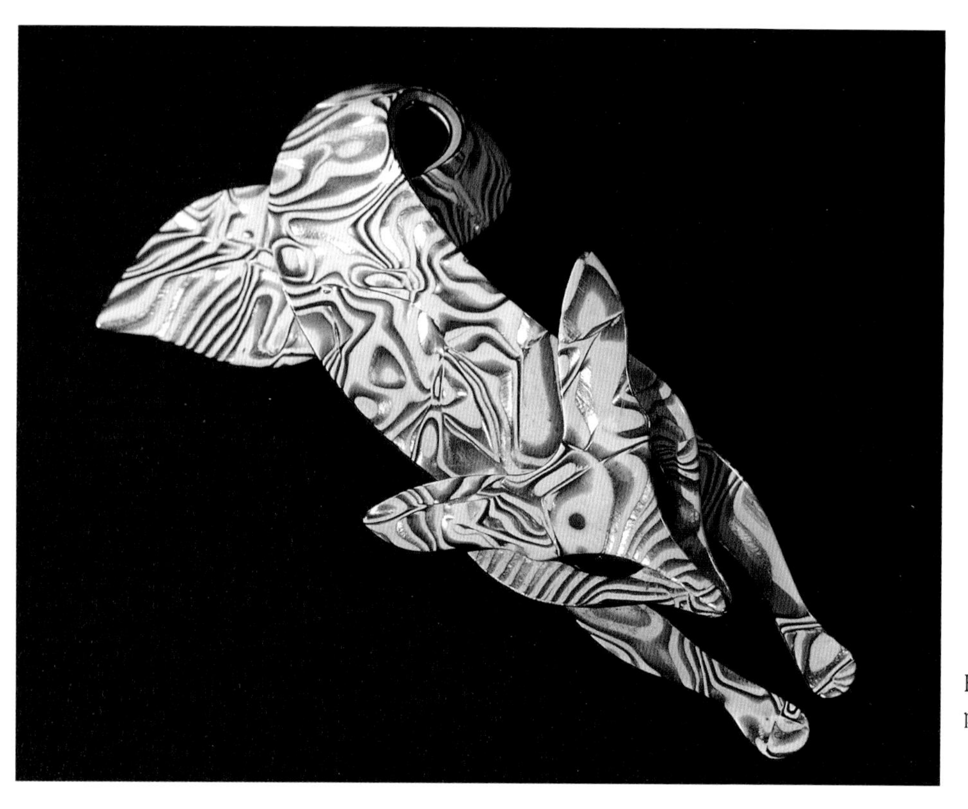

Famous fox pin with layered pattern, c. 1960s to present. 4". *Courtesy of OhMy!!* $125-175.

Fuchsia pearlized famous fox pin, c. 1960s to present. 4". *Courtesy of OhMy!!* $125-175.

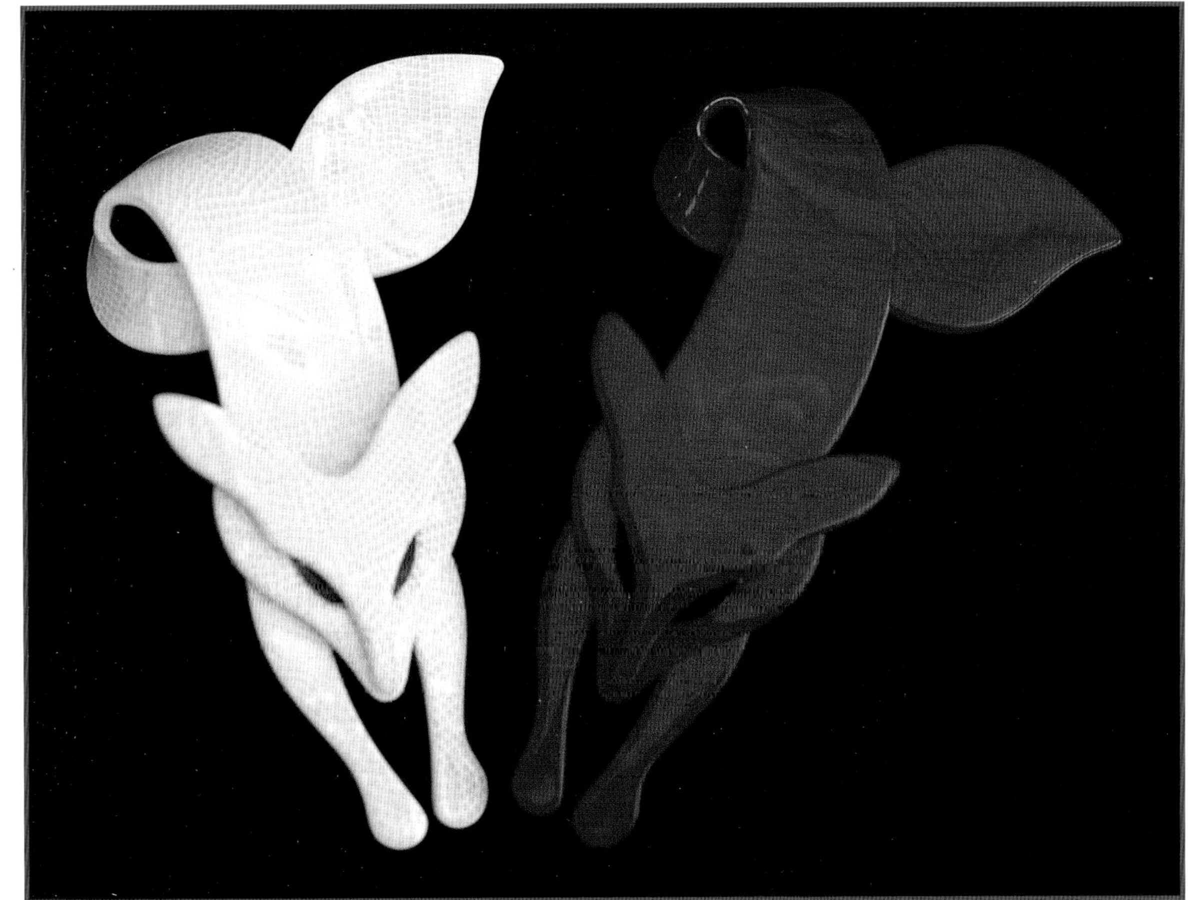

Pair of red and white famous fox pins, c. 1960s to present. 4". *Courtesy of author.* $125-175 each.

Famous fox pins in assorted glitter colors, c. 1960s to present. 4". *Courtesy of Nathalie Bernhard.* $125-175 each.

Famous fox pin in pearlized orange, c. 1960s to present. 4". *Courtesy of author.* $125-175.

Trio of famous fox pins, faux snakeskin, striped effect and solid black with red. c. 1960s to present. 4". *Courtesy of OhMy!!* $125-175 each.

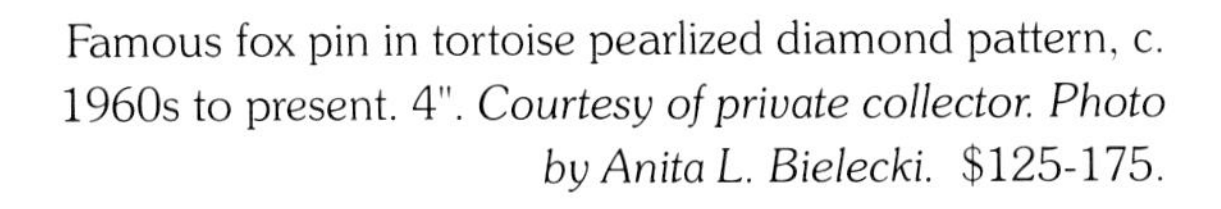

Famous fox pin in tortoise pearlized diamond pattern, c. 1960s to present. 4". *Courtesy of private collector. Photo by Anita L. Bielecki.* $125-175.

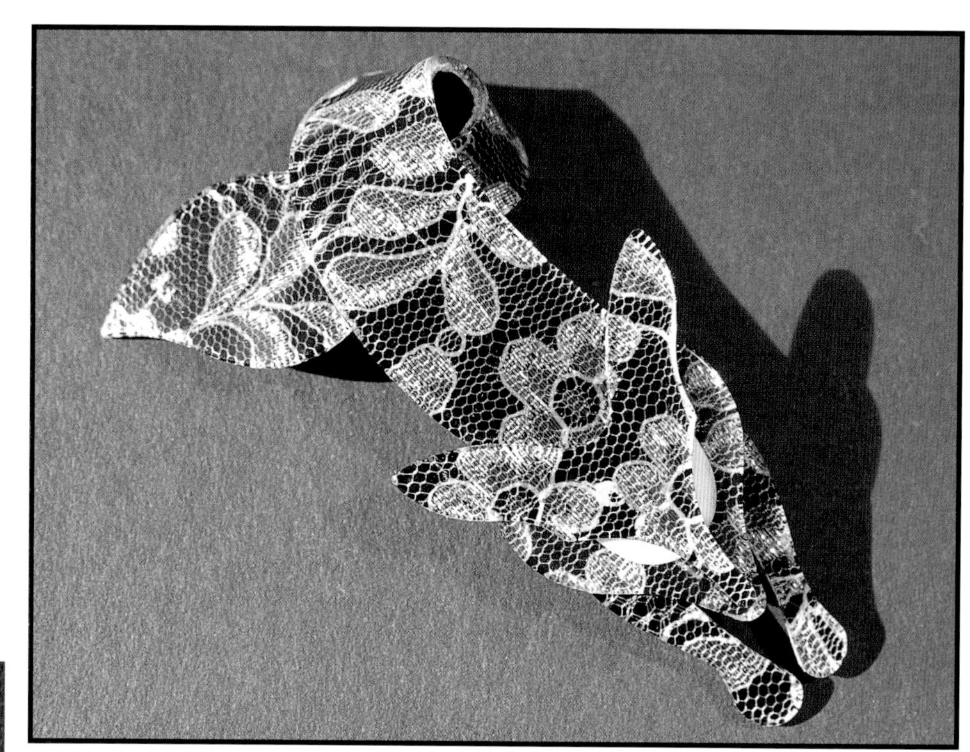

Famous fox pin in lace, gold and black pattern, c. 1960s to present. 4". *Courtesy of OhMy!!* $125-175.

Famous fox pin in marbelized tortoise pattern, c. 1960s to present. 4". *Courtesy of OhMy!!* $125-175.

Chapter 7

Cats and Dogs

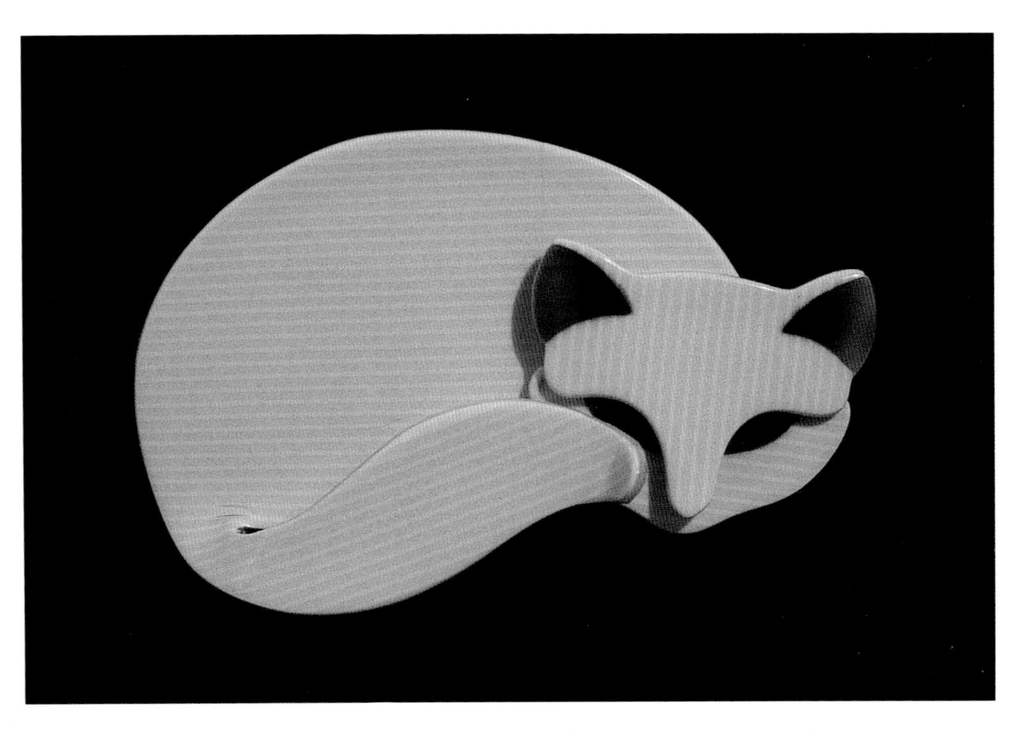

Large sleeping cat pin, ivory with tortoise trim, c. 1990s. 4". *Courtesy of author.* $100-175.

Pair of sleeping cat pins, one ivory with tortoise trim, one all-over tortoise, c. 1990s. *Courtesy of Dale Rhoads and author.* $100-175 each.

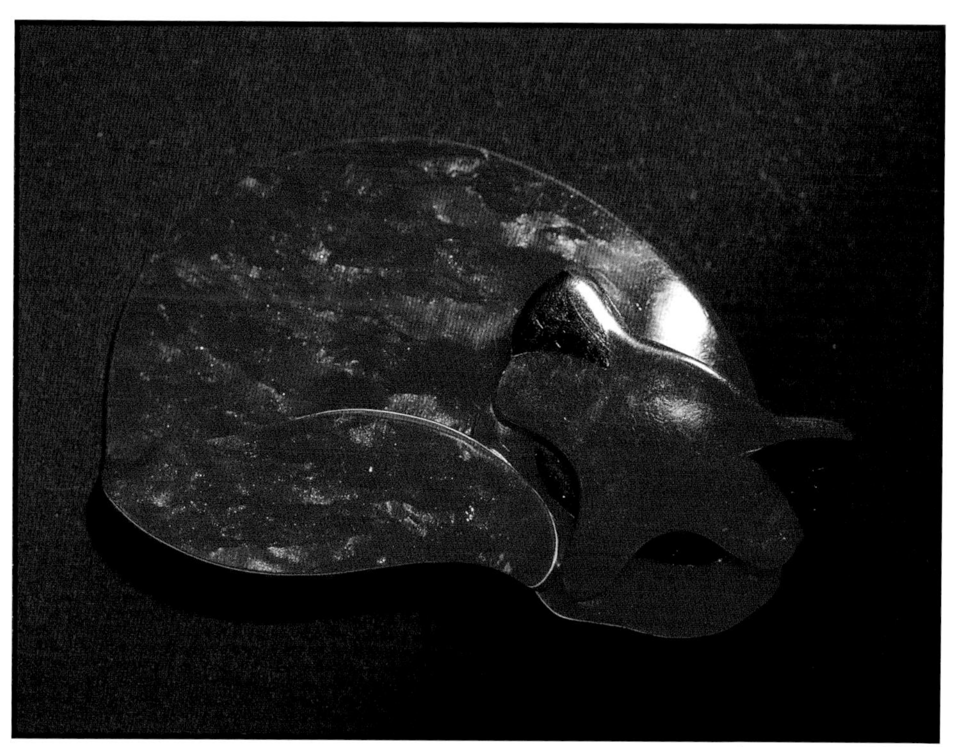
The red sleeping cat pin, c. 1990s. 4". *Courtesy of OhMy!!* $100-175.

The large sleeping cat pin, plaid pattern, c. 1990s. 4". *Courtesy of OhMy!!* $100-175.

Opposite page:
Cat head pins in brown, shaded brown and ivory, stripes, and diamond pattern, c. 1990s. 2 3/8". *Courtesy of OhMy!!* $100-175 each.

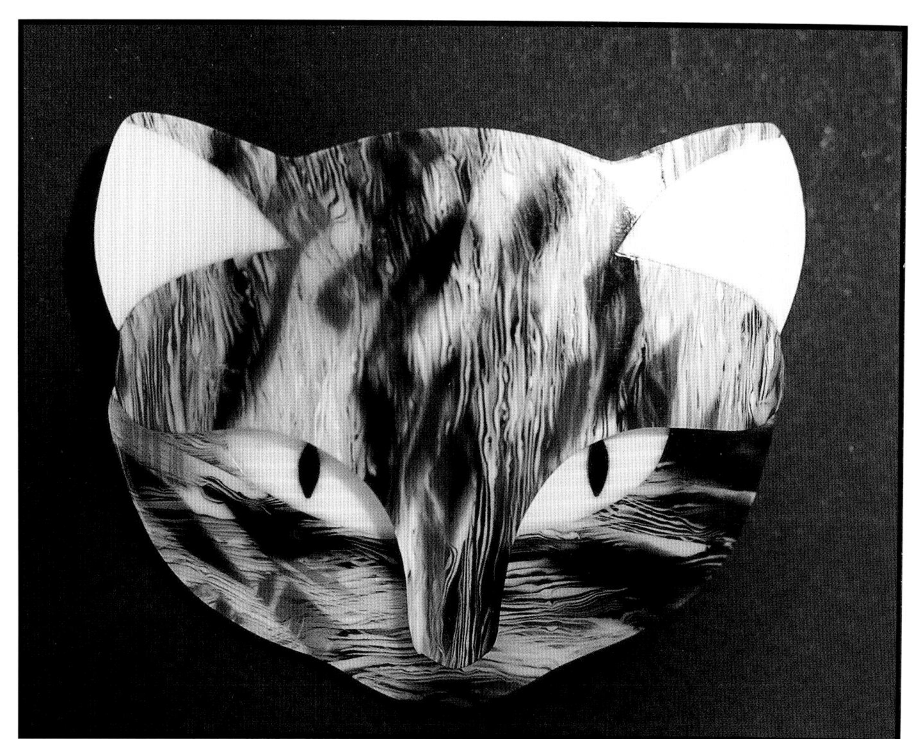

Cat with a ball pin in red and black, c. 1990s. 2 7/8". *Courtesy of OhMy!!* . $100-175.

Felix the cat pin, c. 1968-1980. 2". *Courtesy of OhMy!!* $125-175.

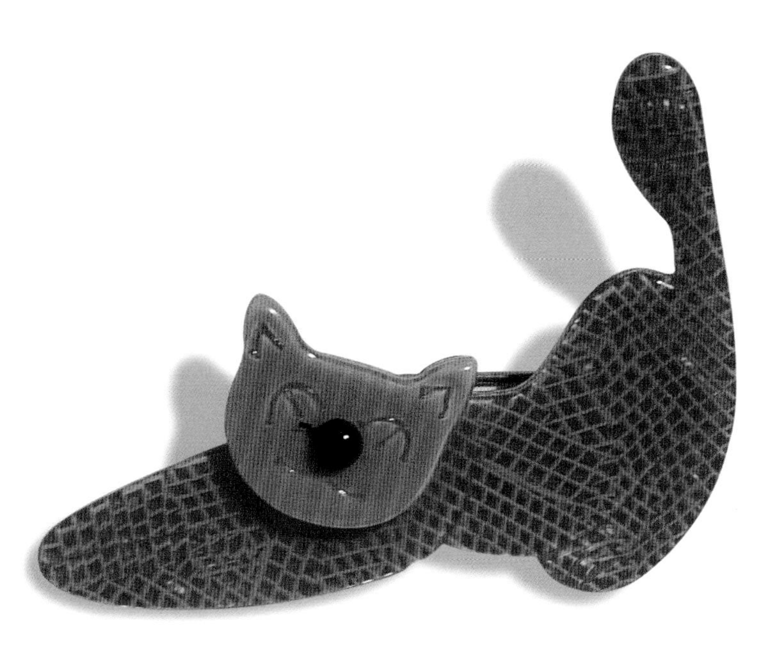

Tomcat pin in pink with black eyes, c. 1968-1980. 2". *Courtesy of Remembrances of Things Past, Provincetown, MA.* $100-175.

Pair of standing cats, one tortoise and one red, c. 1990s. 4 1/8". *Courtesy of OhMy!!* $125-175.

Pair of cats, one with tweed body and red bow, the other with tigereye pearlized body and green bow, c. 1960s. *Courtesy of Nathalie Bernhard.* $150-250 each.

Three standing cats in different patterns, c. 1990s. 4 1/8". *Courtesy of private collector. Photo by Anita L. Bielecki.* $125-175.

Cat with rhinestones and two bird companions, c. 1960s *Courtesy of Lea Stein.* $150-275 each.

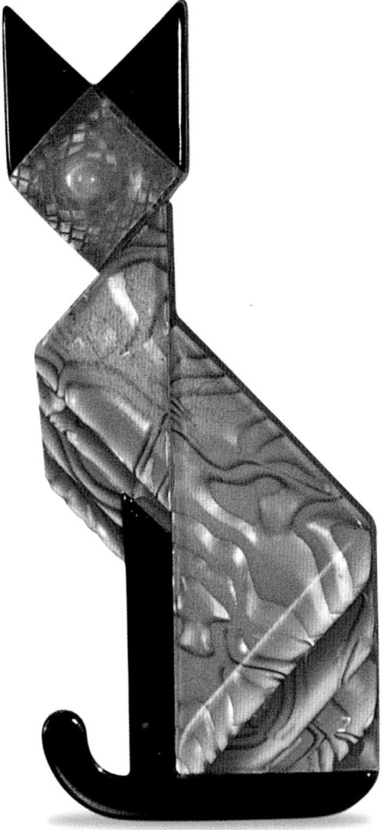

Rare Art Deco style cat with blue swirl pattern, c. 1960s. 2". *Courtesy of Nathalie Bernhard.* $250-375.

Ric the Terrier dog pin with faux mother-of-pearl trim on ears and collar, c. 1988-1990s. 3 5/8". *Courtesy of author.* $100-175.

Pair of Ric the Terrier dog pins, one with mosaic pattern on body and mother-of-pearl trim, the other with stripes and spots and textured trim, c. 1988-1990s. 3 5/8". *Courtesy of private collector. Photo by Anita L. Bielecki.* $100-175 each.

Trio of dachshund pins: one black with rhinestones, one gold pearlized pattern, and one pearlized pink and purple, c. 1968-1980. 3". *Courtesy of Dale Rhoads and author.* $125-175 each.

Ric the Terrier dog pin with black body, c. 1990s. 3 5/8". *Courtesy of OhMy!!.* $100-175.

Poodle pin with ivory pearlized body and tie dyed trim, c. 1968-1980. 1 3/8". *Courtesy of Remembrances of Things Past, Provincetown, MA.* $100-175.

Pair of poodle pins: one black and one red in textured patterns, c. 1968-1980. 1 3/8". *Courtesy of OhMy!! and author.* $125-175.

Pair of Ploukie the dog pins, one is pearlized, the other glitter, c. 1968-1980. 2 1/4". *Courtesy of OhMy!!* $125-175 each.

Red poodle pin with varied textures, c. 1968-1980. 1 3/8". *Courtesy of Remembrances of Things Past, Provincetown, MA.* $125-175.

Totie the scottie pin, red with brown bow, c. 1968-1980. 1 5/8". *Courtesy of OhMy!!* $125-175.

Plouc the dog pin, tortoise and black, c. 1968-1980. 2". *Courtesy of OhMy!!* $125-175.

Totie the scottie pin, multi-glitter with pearlized bow, c. 1968-1980. 1 5/8". *Courtesy of OhMy!!* $125-175.

Chapter 8

More Creatures

Pair of fox head pins: one red and white and the other ivory, black, and tan, c. 1968-1980. 2 3/4". *Courtesy of OhMy!! and author*. $125-175 each.

Four multi-colored fox head pins, c. 1968-1980. 2 3/4". *Courtesy of Remembrances of Things Past, Provincetown, MA.* $125-175 each.

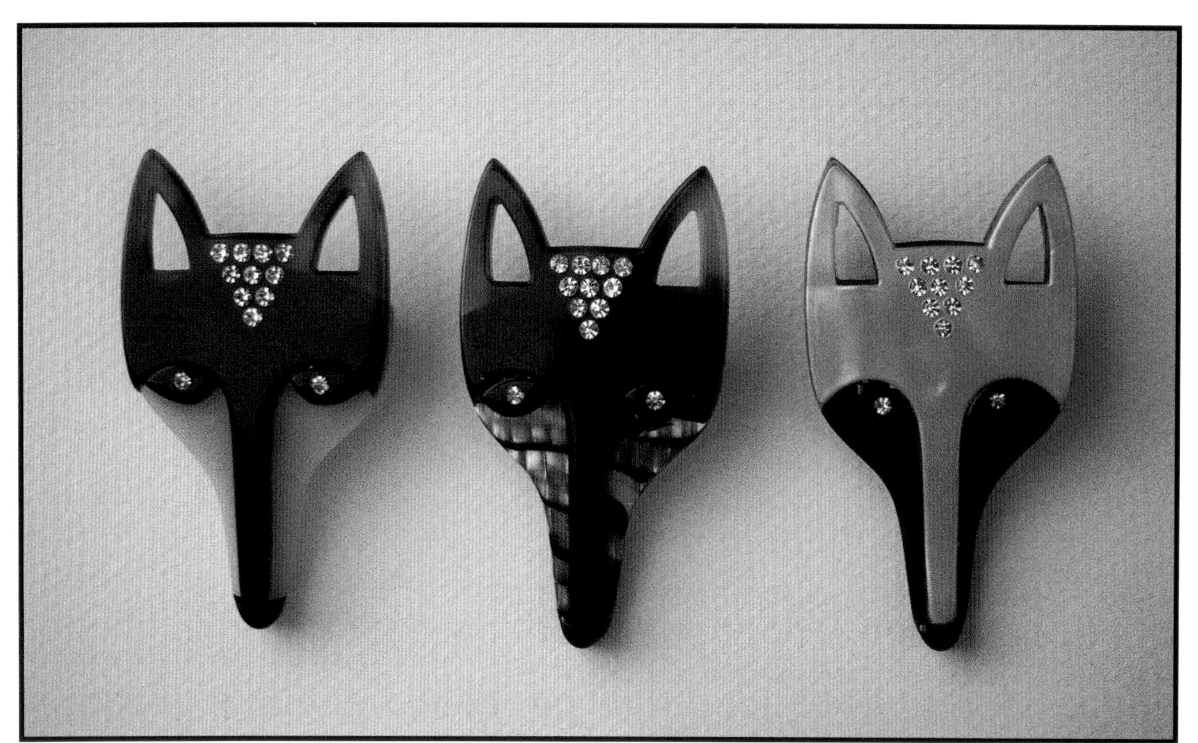

Trio of rhinestone fox head pins in black and ivory, c. 1968-1980. 2 3/4". *Courtesy of Nathalie Bernhard.* $125-175.

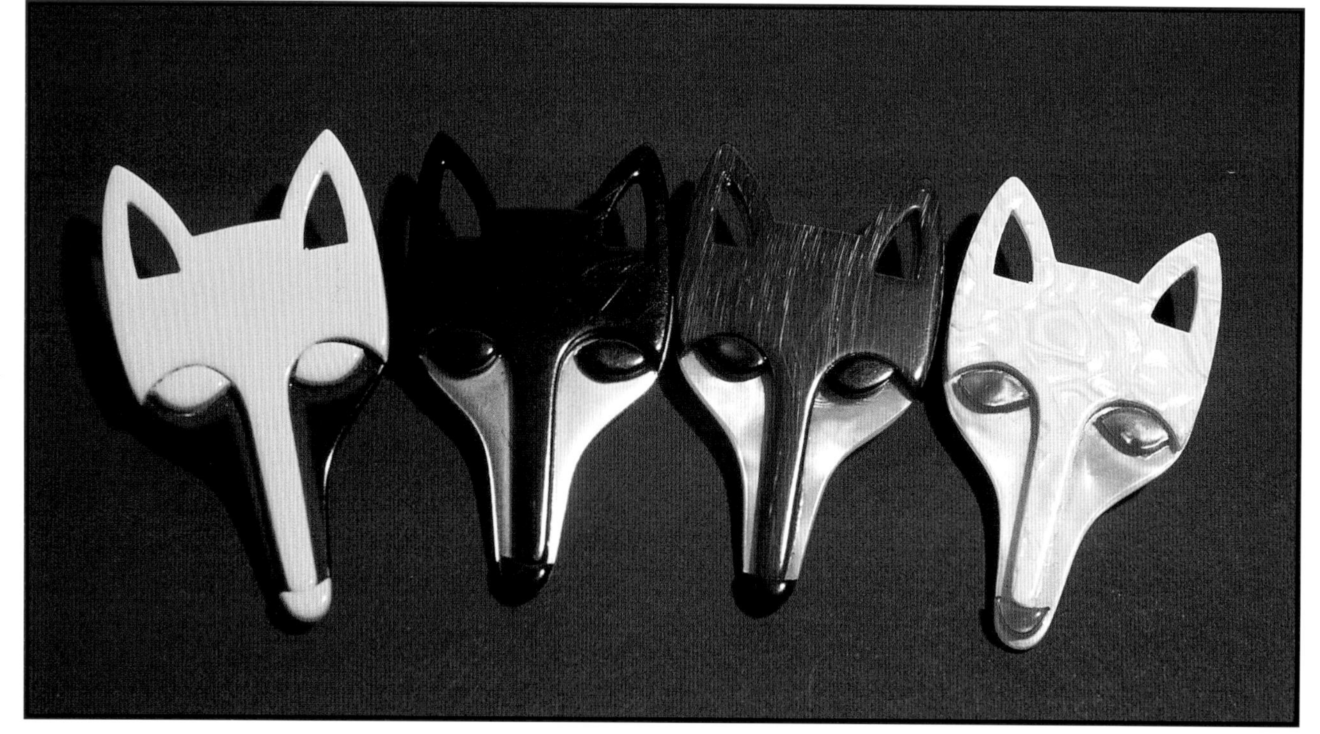

Four fox head pins, all in shades of ivory, tan, and black with pearlized accents, c. 1968-1980. 2 3/4". *Courtesy of Remembrances of Things Past, Provincetown, MA.* $125-175.

Big elephant pin in faux tigereye, c. 1968-1980. 2 3/8". *Courtesy of OhMy!!* $125-175.

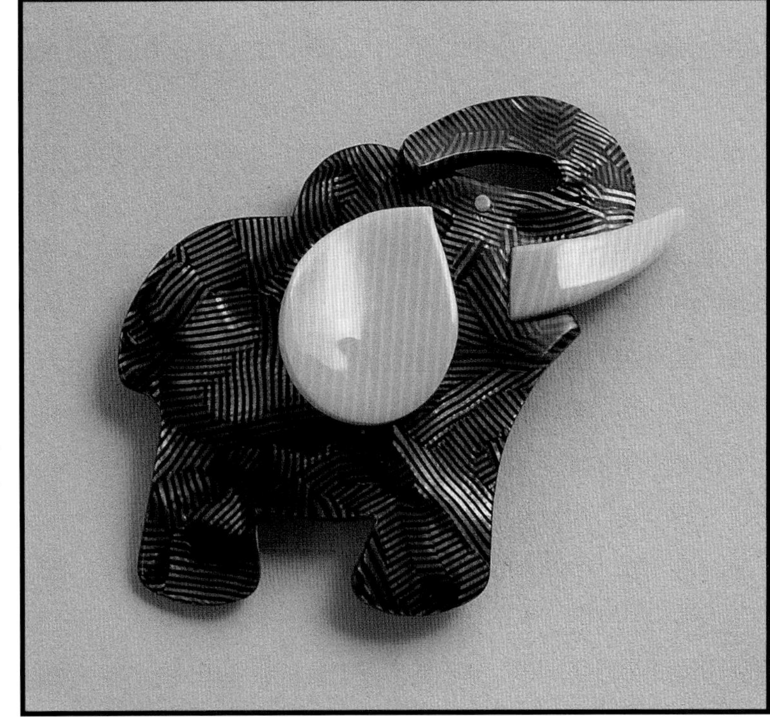

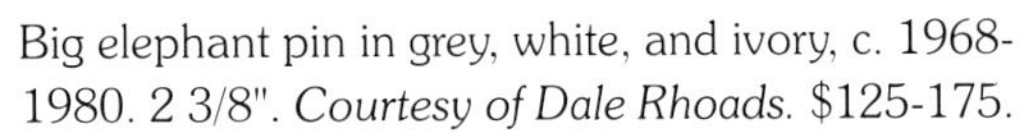

Big elephant pin in grey, white, and ivory, c. 1968-1980. 2 3/8". *Courtesy of Dale Rhoads.* $125-175.

Big elephant pin in ivory and brown, c. 1968-1980. 2 3/8". *Courtesy of Remembrances of Things Past, Provincetown, MA.* $125-175.

Carved red lion pin, c. 1968-1980. 2 1/4". *Courtesy of OhMy!!* $100-175.

Pair of doe pins in black and brown, c. 1968-1980. 2 1/8". *Courtesy of Remembrances of Things Past, Provincetown, MA.* $125-175 each.

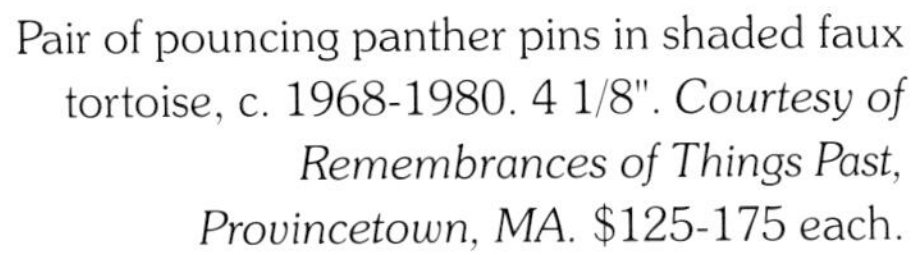

Pair of pouncing panther pins in shaded faux tortoise, c. 1968-1980. 4 1/8". *Courtesy of Remembrances of Things Past, Provincetown, MA.* $125-175 each.

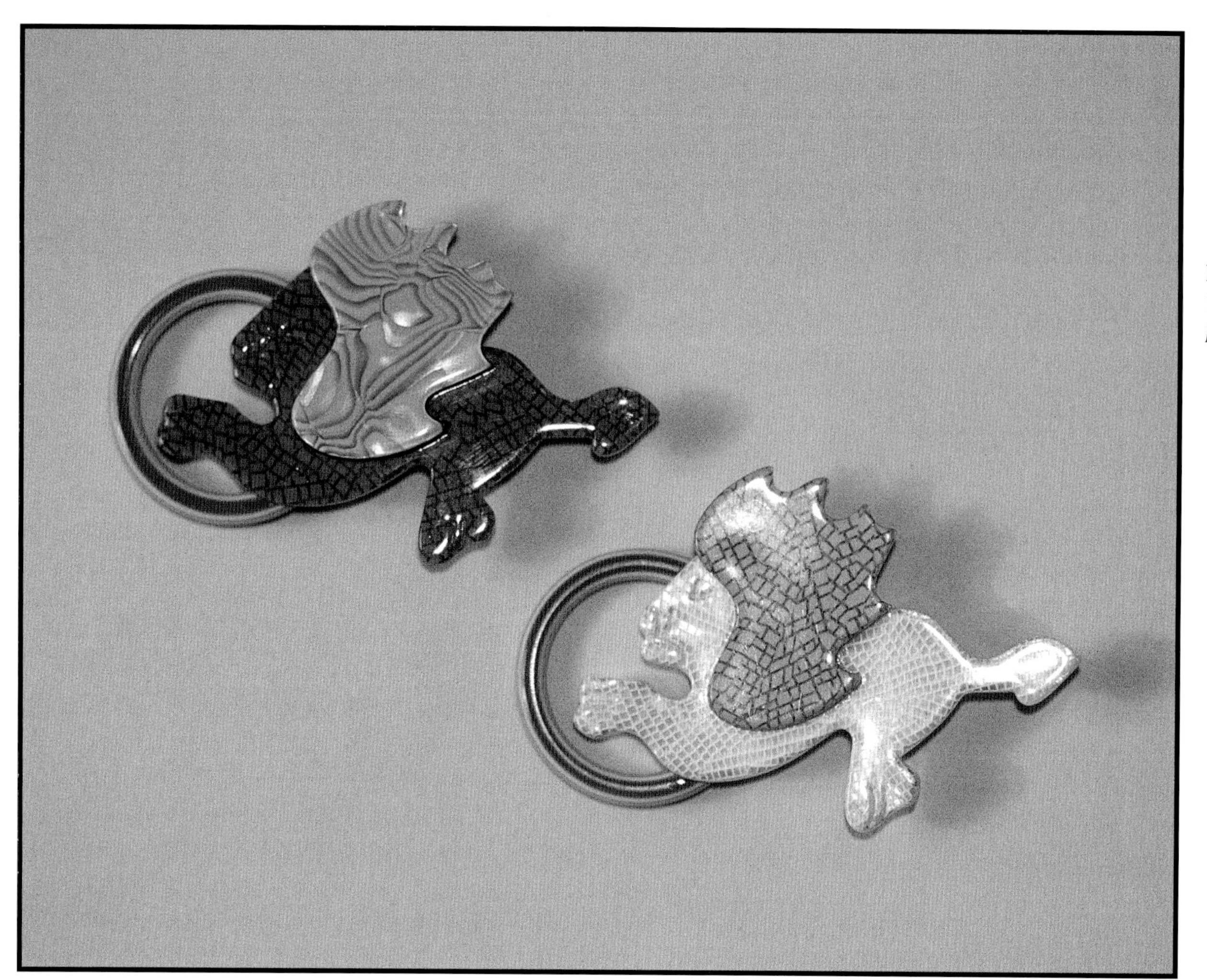

Pair of pearlized lion pins with rings, c. 1968-1980. *Courtesy of Nathalie Bernhard.* $125-175.

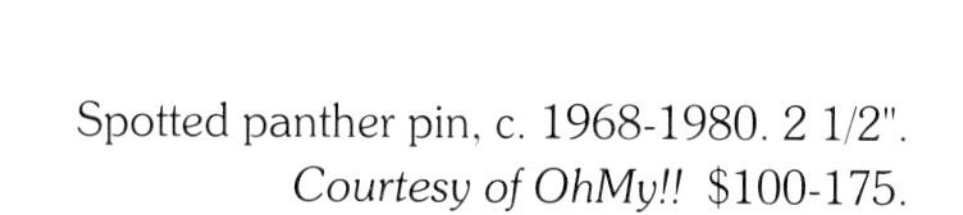

Spotted panther pin, c. 1968-1980. 2 1/2". *Courtesy of OhMy!!* $100-175.

Pair of carved camel pins in ivory and black, c. 1968-1980. 1 3/4". *Courtesy of OhMy!!* $100-175.

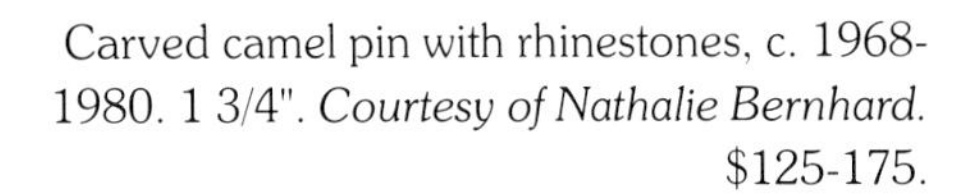

Carved camel pin with rhinestones, c. 1968-1980. 1 3/4". *Courtesy of Nathalie Bernhard.* $125-175.

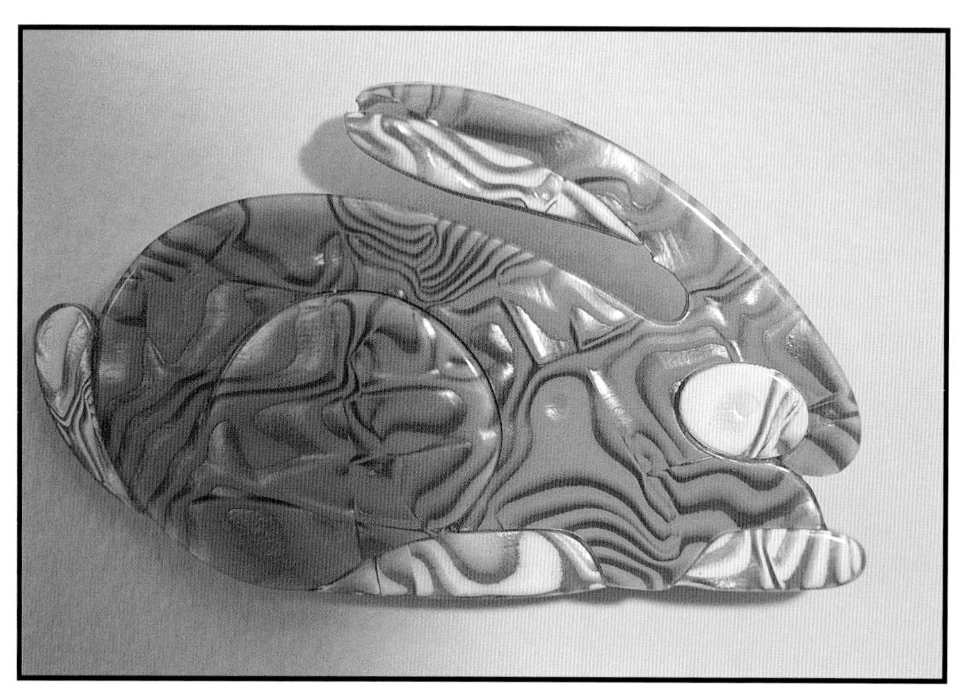

Carved rabbit pin in lavender and white, c. 1968-1980. 2 1/2". *Courtesy of OhMy!!* $125-175.

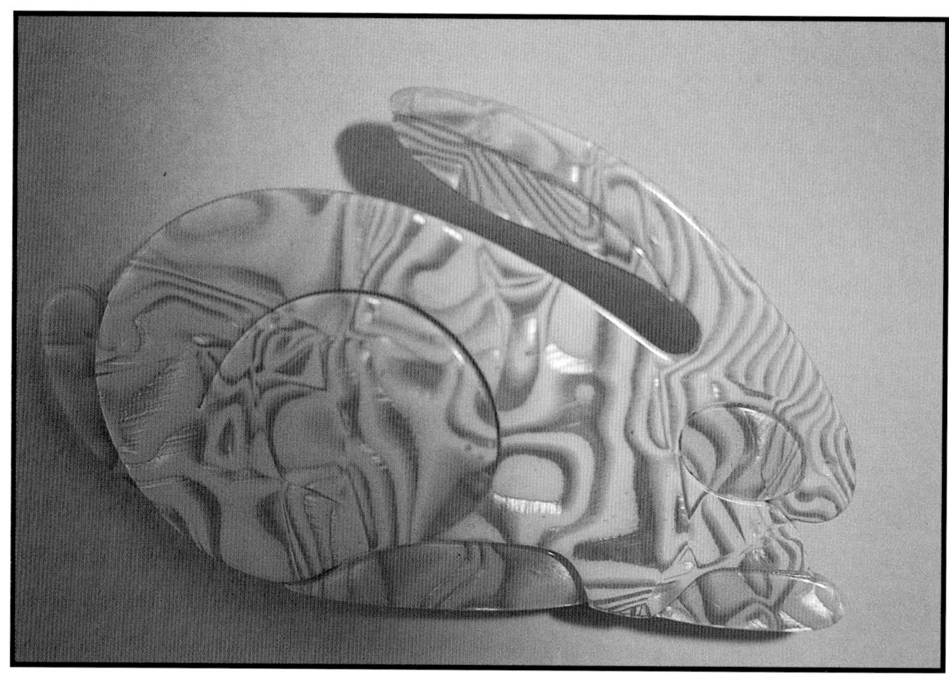

Carved rabbit pin in gold and white, c. 1968-1980. 2 1/2". *Courtesy of author.* $125-175.

Red and multi-colored hippo pin, c. 1968-1980. 1 5/8". *Courtesy of Remembrances of Things Past, Provincetown, MA.* $125-175.

Yellow and multi-colored hippo pin, c. 1968-1980. 1 5/8". *Courtesy of Ohmy!!* $125-175.

Black and multi-colored hippo pin, c. 1968-1980. 1 5/8". *Courtesy of author. $125-175.*

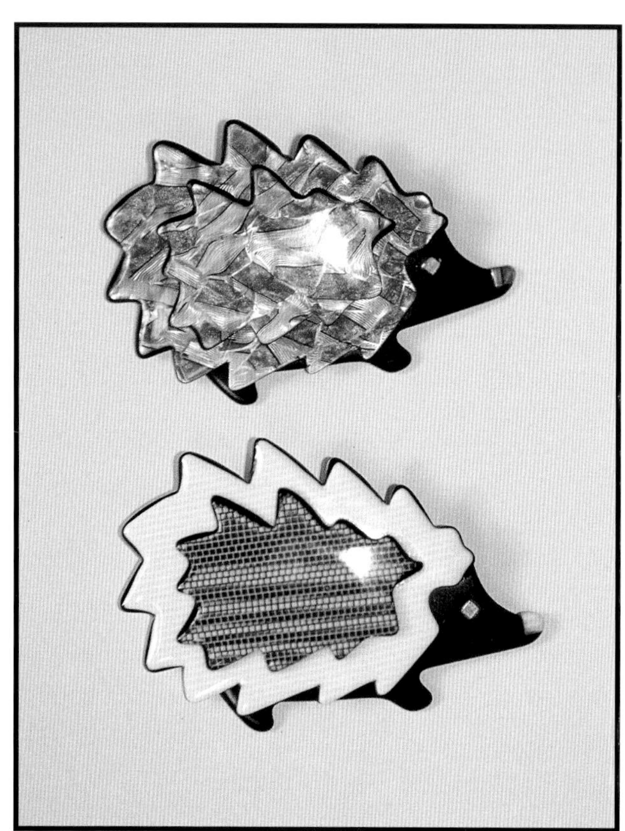

Three pair of porcupine pins in red and black; silver, ivory and black; and tortoise, black, and ivory, c. 2000. One of the first designs for collectors for the new millennium, 3". *Courtesy of Nathalie Bernhard.* $125-175 each.

Alligator pin in tortoise and black, c. 1968-1980. 3 1/4". *Courtesy of Remembrances of Things Past, Provincetown, MA.* $100-175.

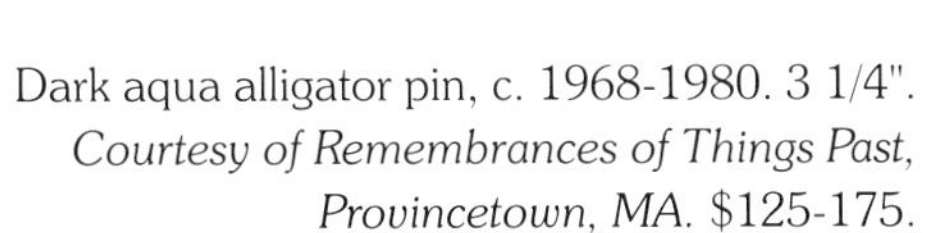

Dark aqua alligator pin, c. 1968-1980. 3 1/4". *Courtesy of Remembrances of Things Past, Provincetown, MA.* $125-175.

Rare koala bear pin, c. 1968-1980.
Courtesy of Lea Stein Collection, Paris.

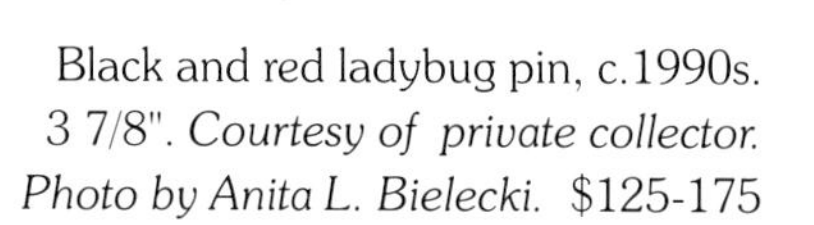

Black and red ladybug pin, c.1990s.
3 7/8". *Courtesy of private collector. Photo by Anita L. Bielecki.* $125-175

Four tortoise pins in mosaic, multi-color tortoise, gold glitter and black, and silver glitter and black, c. 1990s. 3 3/8". *Courtesy of OhMy!!* $125-175 each.

Mult-colored fish pin, c. 1968-1980. 2 3/8" *Courtesy of OhMy!!* $100-175.

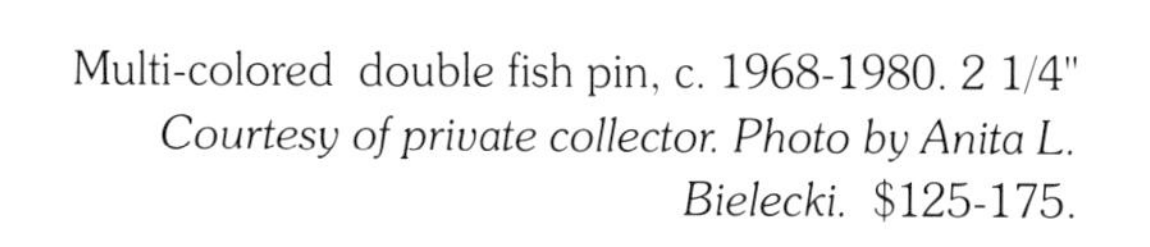

Multi-colored double fish pin, c. 1968-1980. 2 1/4" *Courtesy of private collector. Photo by Anita L. Bielecki.* $125-175.

Chapter 9

Birds and the Bees

Four multi-colored owl pins, c. 1990s. 2 5/8". *Courtesy of OhMy!! and author.* $125-175 each.

Pearlized owl pin, c. 1990s. 2 5/8".
Courtesy of author. $125-175.

Gold and orange hoot owl pin, c. 1970s, 2 1/4". *Courtesy of Dale Rhoads.* $125-195.

Large, multi-colored mosaic owl head pin, c. 1990s. 3 7/8". *Courtesy of OhMy!!* $125-175.

Trio of large, multi-patterned and multi-colored owl pins, c. 1990s. 3 7/8" *Courtesy of OhMy!!* $125-175 each.

A three ducks pin, c. 1968-1980. 2 1/8". *Courtesy of Remembrances of Things Past, Provincetown, MA.* $125-175.

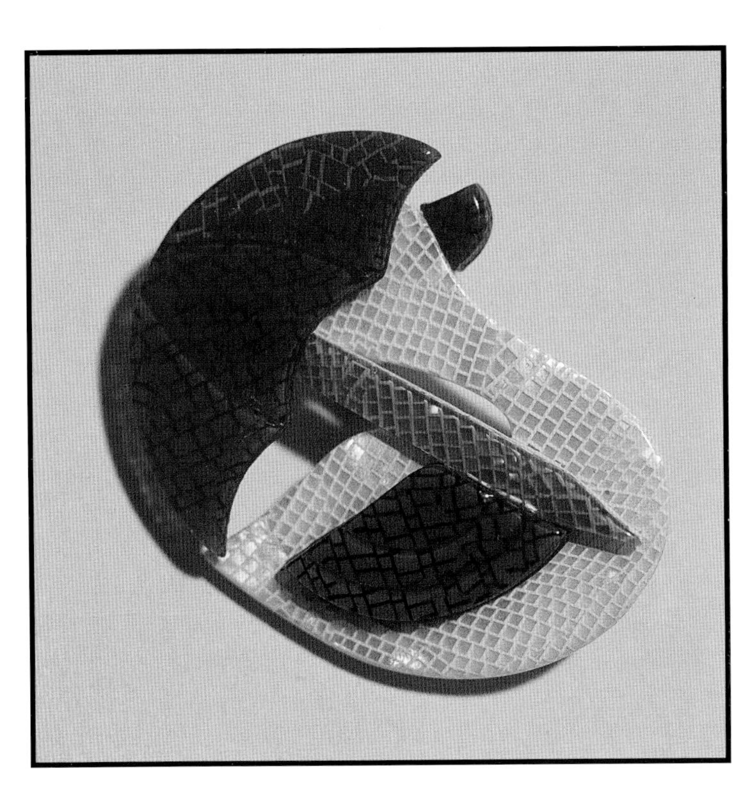

Multi-colored duck with umbrella pin, c. 1968-1980. 1 5/8". *Courtesy of OhMy!!* $125-175.

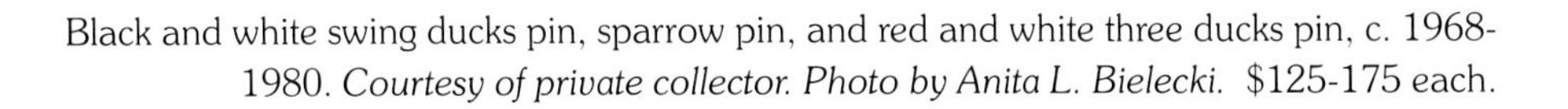

Black and white swing ducks pin, sparrow pin, and red and white three ducks pin, c. 1968-1980. *Courtesy of private collector. Photo by Anita L. Bielecki.* $125-175 each.

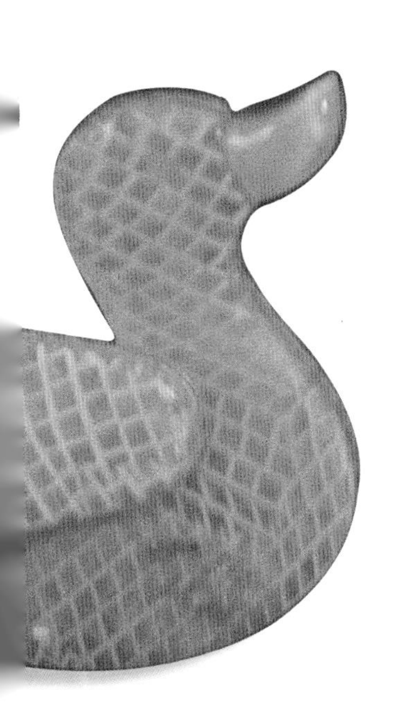

Gold exotic bird pin, c. 1968-1980. 1 5/8". *Courtesy of OhMy!!* $125-175.

"Ying and Yang" bird pins in multi-colors, c. 1968-1980. *Courtesy of Nathalie Bernhard.* $125-175 each.

Trio of great beak bird pins in multi-colors, c. 1968-1980. 1 7/8". *Courtesy of OhMy!!* $125-175 each.

Black and white great beak bird pin, c. 1968-1980. 1 7/8". *Courtesy of private collector. Photo by Anita L. Bielecki.* $125-175.

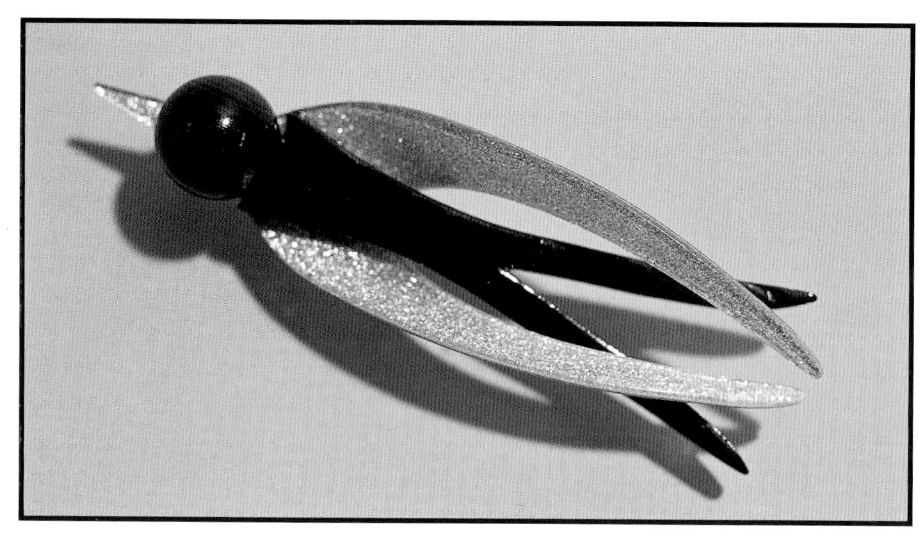

Black and silver glitter swallow pin, c. 1968-1980. 3 3/8" *Courtesy of OhMy!!* $125-175.

Red and silver glitter swallow pin, c. 1968-1980. 3 3/8" *Courtesy of Remembrances of Things Past, Provincetown, MA.* $125-175.

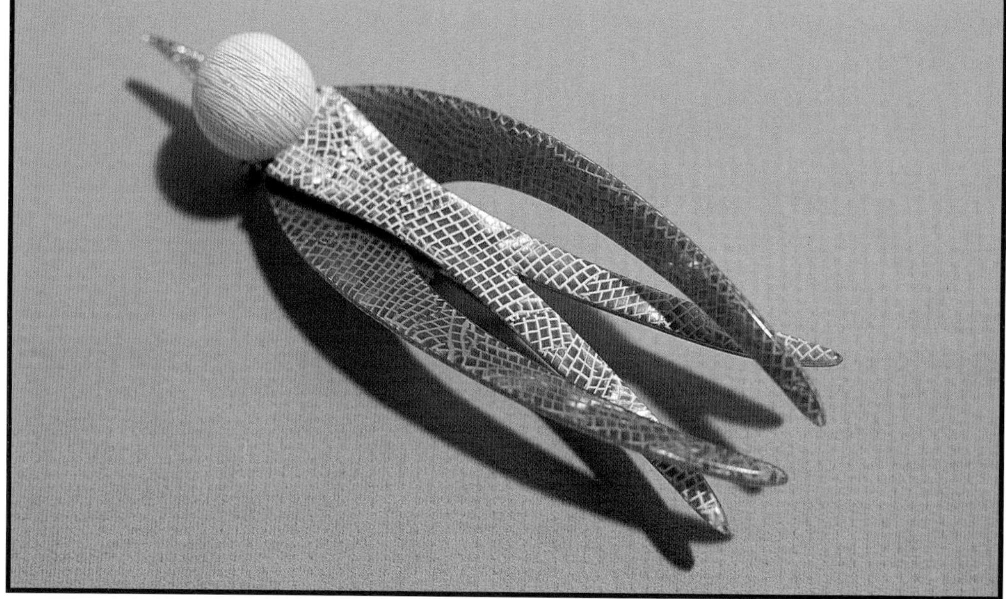

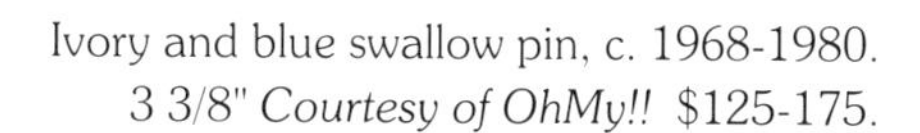

Ivory and blue swallow pin, c. 1968-1980. 3 3/8" *Courtesy of OhMy!!* $125-175.

Red and ivory double sparrow pin, c. 1968-1980. 2 5/8". *Courtesy of Remembrances of Things Past, Provincetown, MA.* $125-175.

Red and black sparrow pin, c. 1968-1980. 1 7/8". *Courtesy of Remembrances of Things Past, Provincetown, MA.* $125-175.

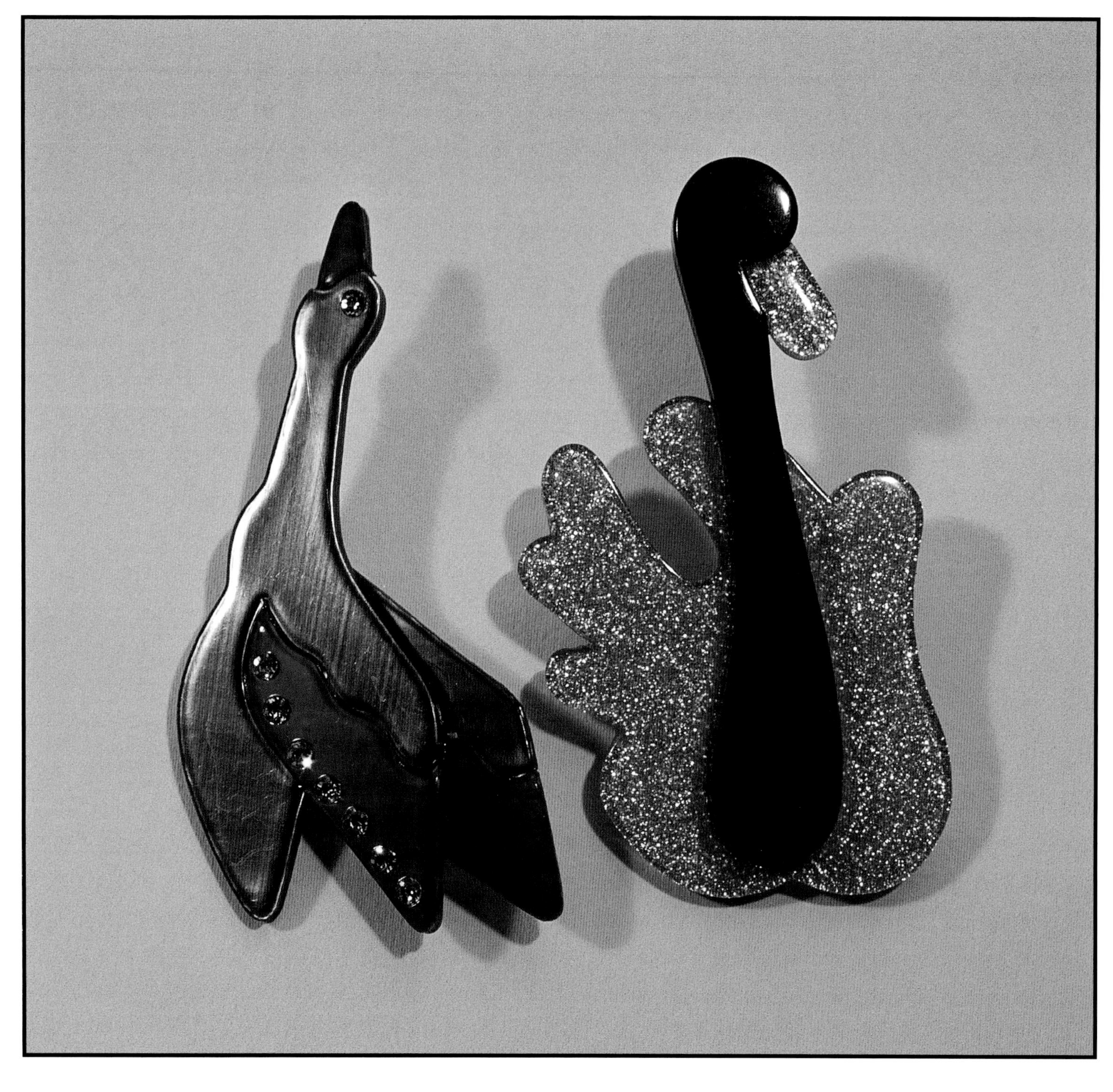

Goose with rhinestones pin and gold glitter swan pin, c. 1968-1980. 2 3/4". *Courtesy of Dale Rhoads.* $125-200 each.

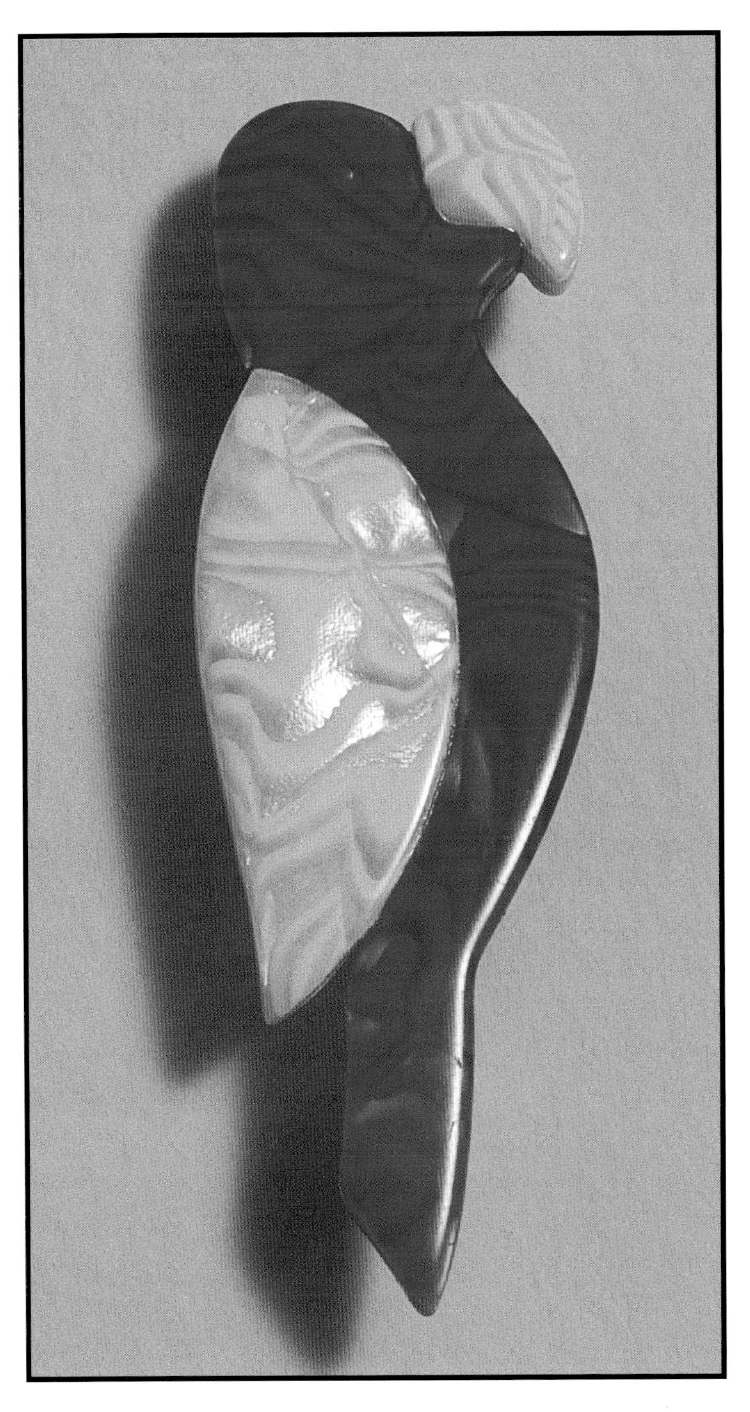

Multi-colored and gold parrot pins, c. 1968-1980. 3 1/2". *Courtesy of Remembrances of Things Past, Provincetown, MA.* $125-175 each.

Trio of parrot pins in multi-colors, c. 1968-1980. 3 1/2". *Courtesy of OhMy!!* $125-175 each.

Ivory and black parrot pin with rhinestones, c. 1968-1980. 3 1/2".
Courtesy of private collector. Photo by Anita L. Bielecki. $125-175.

Black and ivory parrot pin with green rhinestones, c. 1968-1980. 3 1/2". *Courtesy of Nathalie Bernhard.* $125-175.

Pearlized flamingo pin, c. 1968-1980. 2 5/8". *Courtesy of OhMy!!* $125-175.

Pair of multi-colored walking geese pins, c. 1968-1980. 2". *Courtesy of Nathalie Bernhard.* $125-175 each.

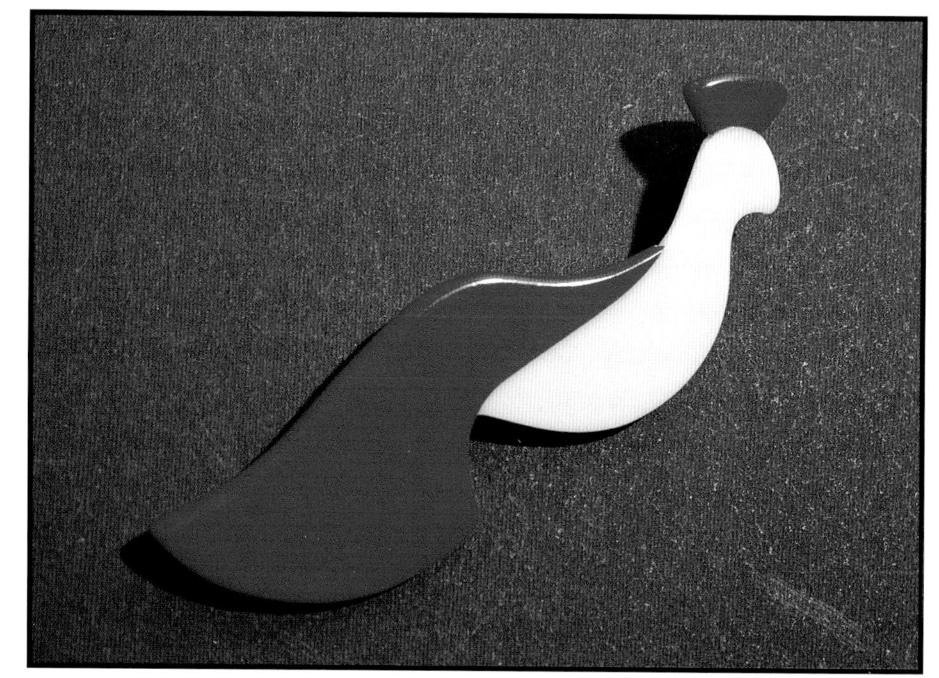

Red and ivory peacock pin, c. 1968-1980. 3". *Courtesy of Dale Rhoads.* $125-175.

Red and multi-colored toucan bird pin, c. 1968-1980. 2 1/8". *Courtesy of Remembrances of Things Past, Provincetown, MA.* $125-175.

Black and multi-colored toucan bird pin, c. 1968-1980. 2 1/8". *Courtesy of OhMy!!* $125-175.

Gold and ivory striped bee pin with gold wings, c. 1968-1980. 2 3/8". *Courtesy of author.* $125-175.

Purple and multi-colored penguin pin, c. 1968-1980. 2 1/2". *Courtesy of OhMy!!* $125-175.

Black and ivory striped bee pin with gold and black wings, c. 1968-1980. 2 3/8". *Courtesy of OhMy!!* $125-175.

Green striped bee pin with gold wings, c. 1968-1980. 2 3/8". *Courtesy of private collector. Photo by Anita L. Bielecki.* $125-175.

Red and ivory stripe bee pin with red and gold wings, c. 1968-1980. 2 3/8". *Courtesy of OhMy!!* $125-175.

Chapter 10

Butterflies and Flowers

Red and ivory butterfly pin, c. 1968-1980. 2". *Courtesy of Remembrances of Things Past, Provincetown, MA.* $125-175.

Pair of pearlized butterfly pins: one with blue rhinestones, c. 1968-1980. 2". *Courtesy of OhMy!! and author.* $125-175 each.

Trio of multi-colored butterfly pins, c. 1968-1980. 2". *Courtesy of Nathalie Bernhard.* $125-175.

Pair of multi-colored butterfly pins, c. 1968-1980. *Courtesy of private collector. Photo by Anita L. Bielecki.* $125-175 each.

Trio of glitter and pearlized, multi-colored butterfly pins, c. 1968-1980. *Courtesy of OhMy!!* $125-175.

Pair of tiny black butterfly pins with gold insets, c. 1968-1980. approx. 1/2". *Courtesy of Nathalie Bernhard.* $100-150.

Black and tan striped butterfly pin, c. 1960-1980.
Courtesy of Remembrances of Things Past, Provincetown, MA. $100-150.

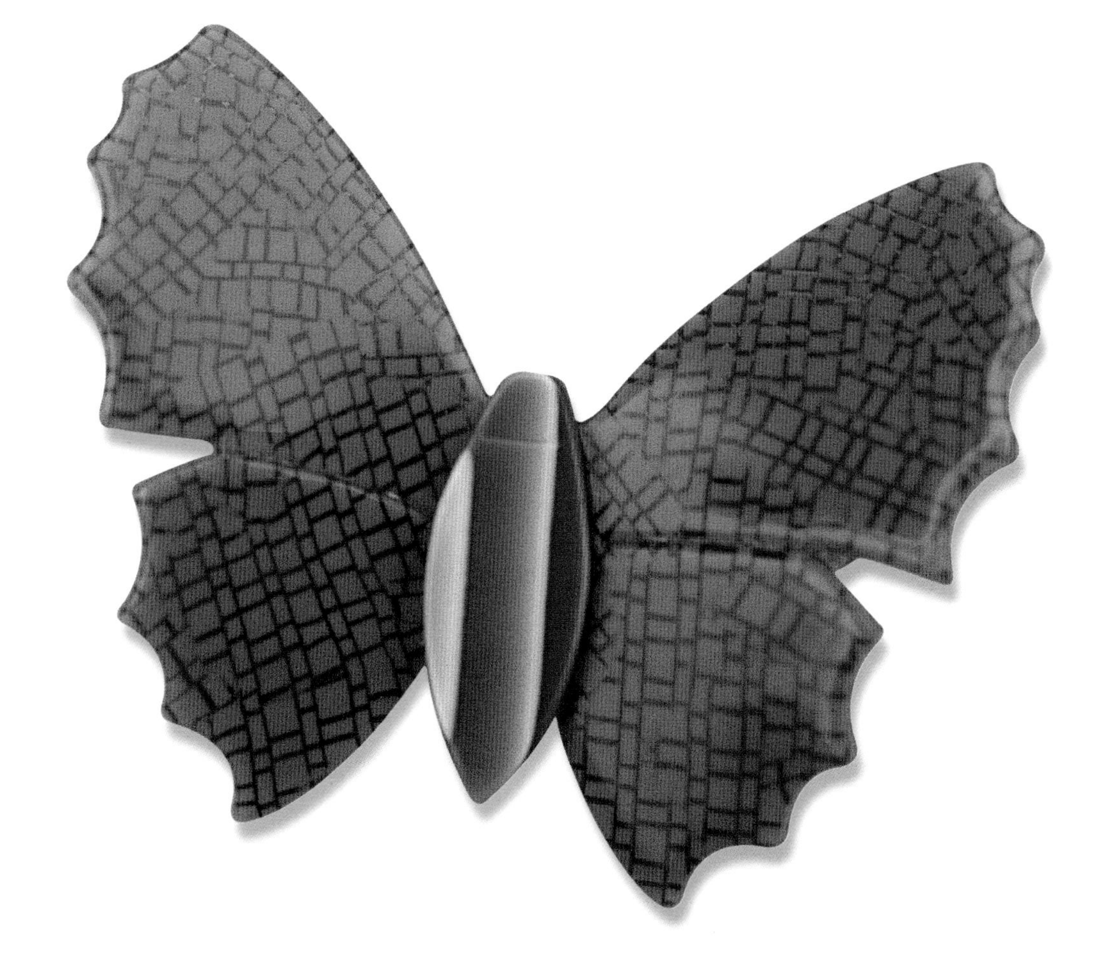

Purple butterfly pin, c. 1968-1980.
Courtesy of Dale Rhoads. $125-175.

Red and black blueberry pin, c. 1968-1980. 2 7/8". *Courtesy of OhMy!!* $125-175.

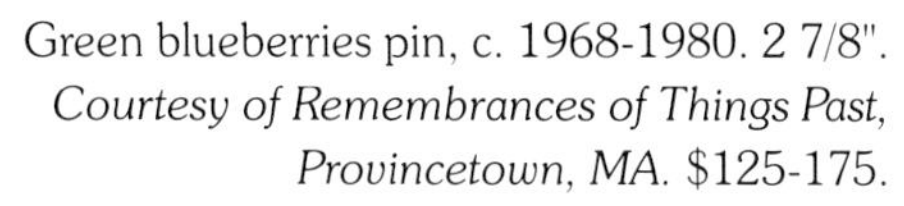

Green blueberries pin, c. 1968-1980. 2 7/8". *Courtesy of Remembrances of Things Past, Provincetown, MA.* $125-175.

Black and white blueberries pin, c. 1968-1980. 2 7/8". *Courtesy of private collector. Photo by Anita L. Bielecki.* $125-175.

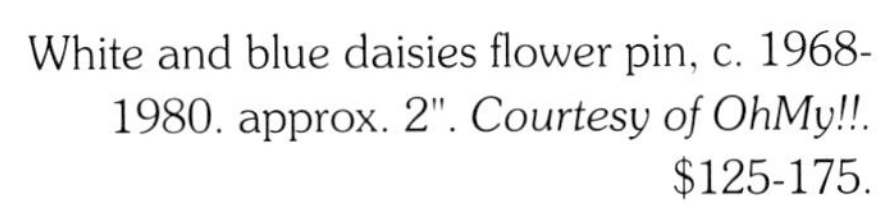

White and blue daisies flower pin, c. 1968-1980. approx. 2". *Courtesy of OhMy!!.* $125-175.

Red and white daisies flower pin, c. 1968-1980. approx. 2". *Courtesy of private collector. Photo by Anita L. Bielecki.* $125-175.

Silver glitter and black three flowers pin, c. 1968-1980. 2 1/8". *Courtesy of OhMy!! $125-175.*

Trio of multi-colored leaf pins, c. 1990-2000. *Courtesy of Nathalie Bernhard.* $125-175 each.

Pearlized carved triple leaves pin, c. 1968-1980. *Courtesy of private collector. Photo by Anita L. Bielecki.* $125-175.

Two engraved pearlized leaf pins: red and brown, c. 1968-1980. 5 7/8". *Courtesy of OhMy!!* $125-175 each.

Multi-colored edelweiss flower pin, c. 1968-1960. 3 3/8". *Courtesy of OhMy!!* $125-175.

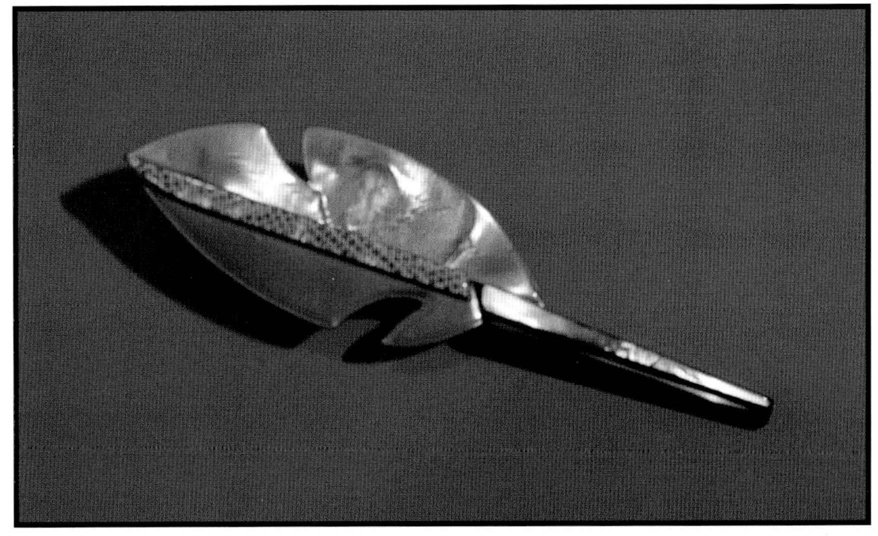

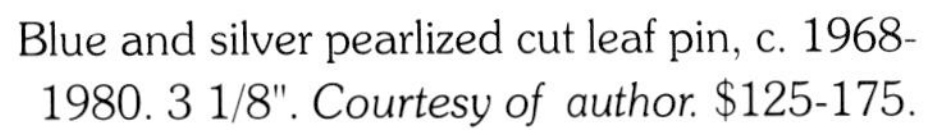

Blue and silver pearlized cut leaf pin, c. 1968-1980. 3 1/8". *Courtesy of author.* $125-175.

Chapter 11

Many Splendid Things

Red pin with hat and straight handle walking stick, c. 1968-1980. 3". *Courtesy of author.* $125-175

Pair of pins with hat and curved handle walking stick, c. 1968-1980. 3". *Courtesy of OhMy!! and author.* $125-175 each.

Two pins in the shapes of hats, c. 1968-1980. *Courtesy of OhMy!!* $125-175.

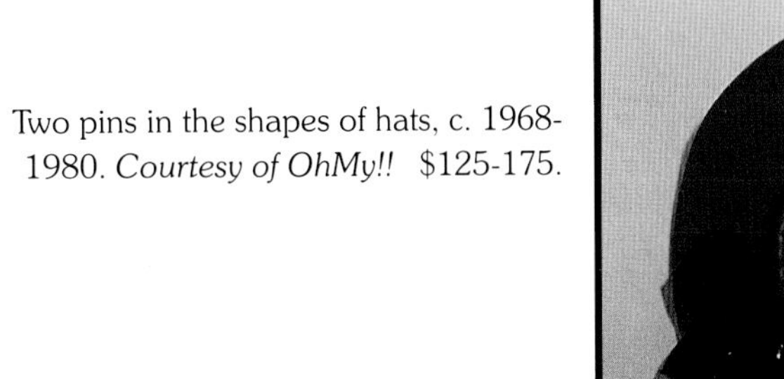

Pair of pins as straw hats, one brown with bow, other tan with bow, c. 1968-1980. *Courtesy of OhMy!!* $125-175.

Blue translucent pin with hat and curved handle walking stick, c. 1968-1980. *Courtesy of author.* $125-175.

Pair of pins shaped as hats with curved walking stick handles: translucent yellow, and red c. 1968-1980. 3". *Courtesy of private collector. Photo by Anita L. Bielecki.* $125-175.

Translucent tan pin shaped as a hat, c. 1968-1980. *Courtesy of Remembrances of Things Past, Provincetown, MA.* $125-175.

Pair of heart pins with arrows and rhinestones, blue and red, c. 1968-1980. *Courtesy of Nathalie Bernhard.* $125-175.

Heart pin with rhinestones and matching round pierced earrings, c. 1968-1980. *Courtesy of private collector. Photo by Anita L. Bielecki.* $250-350 set.

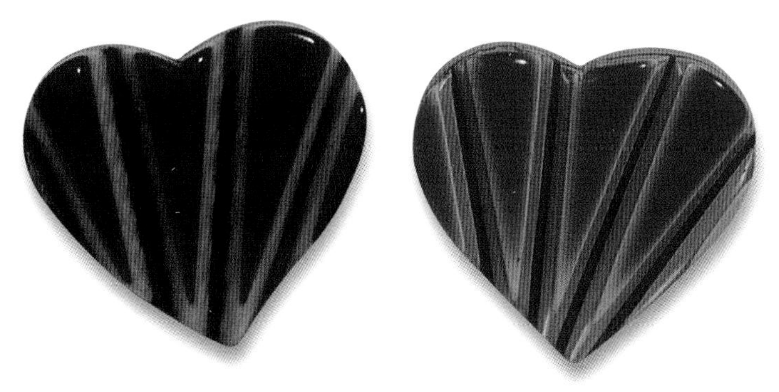

Two engraved and carved heart pins, c. 1968-1980. approx. 3/4". *Courtesy of Remembrances of Things Past, Provincetown, MA.* $125-175 each.

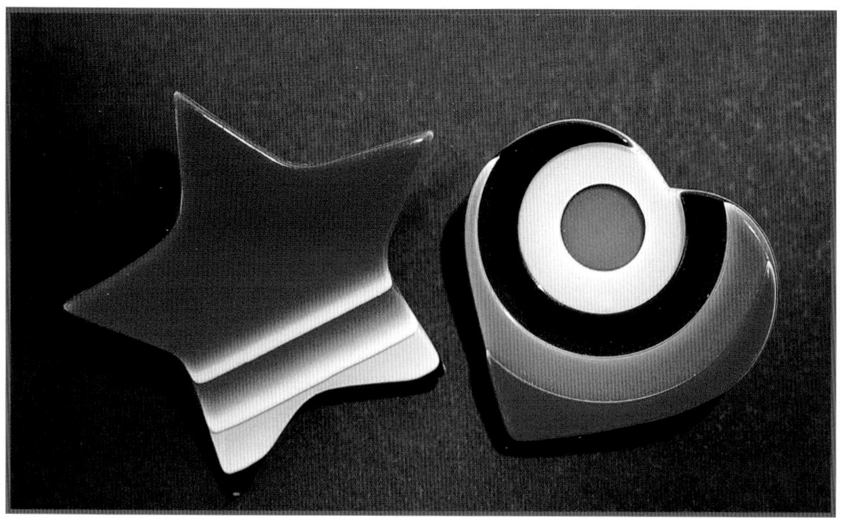

Heart and star pins with multi-colored layers, c. 1968-1980. 1 1/2". *Courtesy private collector.* $125-175 each.

Silver glitter heart pin and two star pins of glitter and black with rhinestones, c. 1968-1980. *Courtesy of OhMy!!* $100-175 each.

Large glitter star pin, small glitter star pin with rhinestones, heart glitter pin, c. !968-1980. *Courtesy of OhMy!!* $125-175 each.

Two comet pins with rhinestones, black and red, c. 1968-1980. 2 5/8". *Courtesy of OhMy!!* $125-175 each.

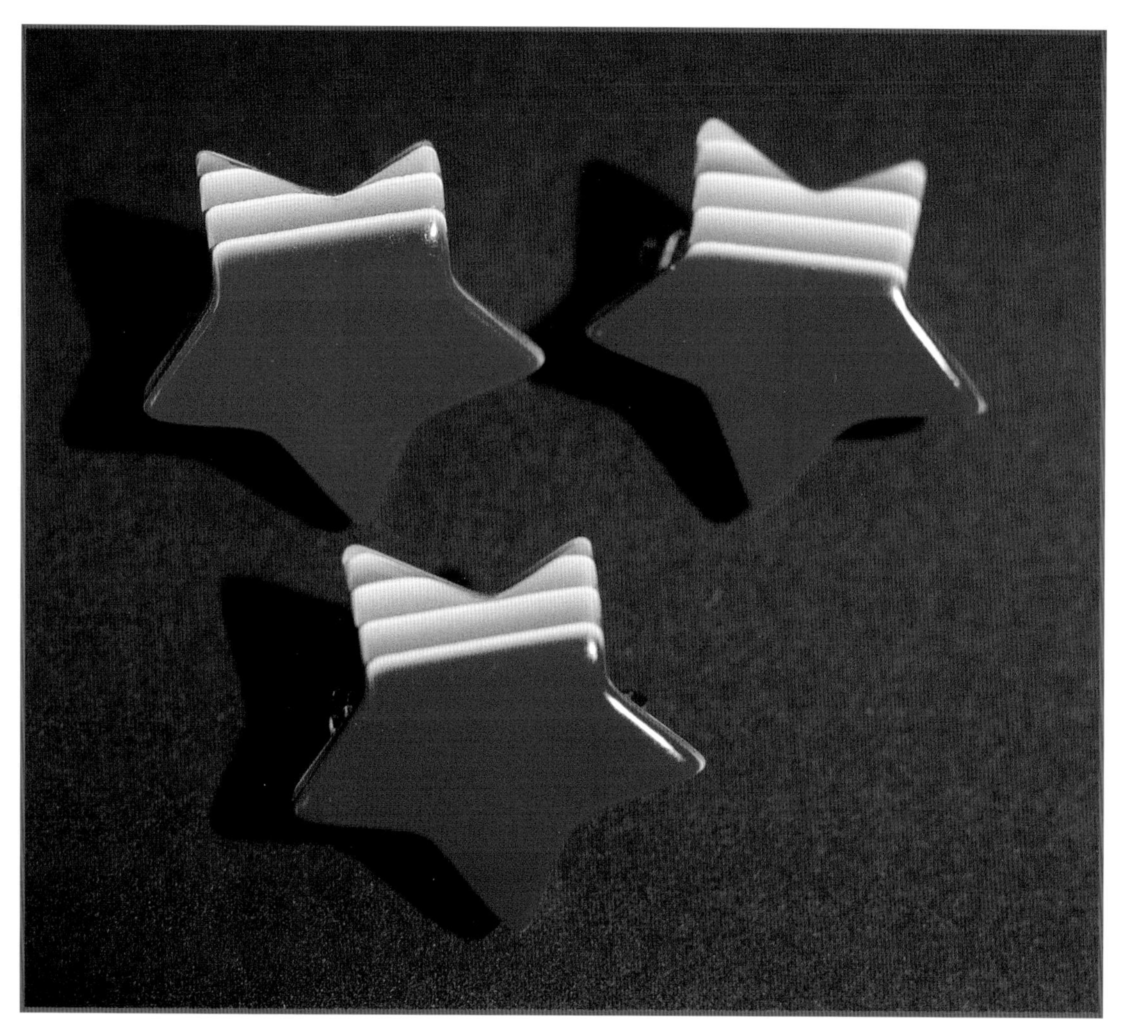

Trio of tiny red layered star pins, c. 1968-1980. 3/4". *Courtesy of author.* $125-175 set.

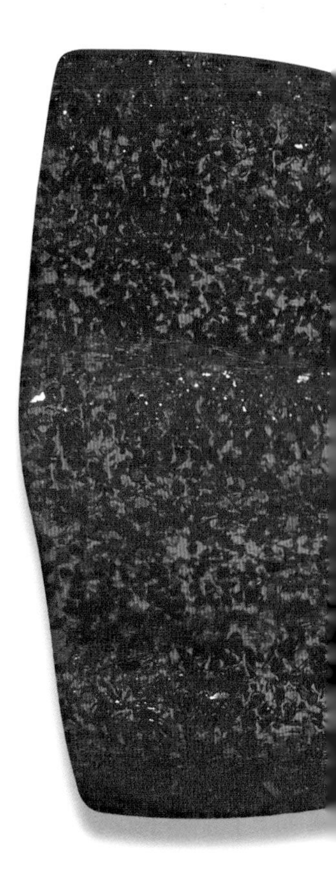

Pair of dimensional star pins, red and blue, c. 1968-1980. *Courtesy of Nathalie Bernhard.* $125-175 each.

Red glitter bow tie pin, c. 1968-1980. 3". *Courtesy of OhMy!!* $125-175.

Burgundy plaid bow tie pin, c. 1968-1980. 2 1/4" *Courtesy of author.* $125-175.

Black and silver pearlized bow tie pin, c. 1968-1980. 2 1/4". *Courtesy of OhMy!!* $100-175.

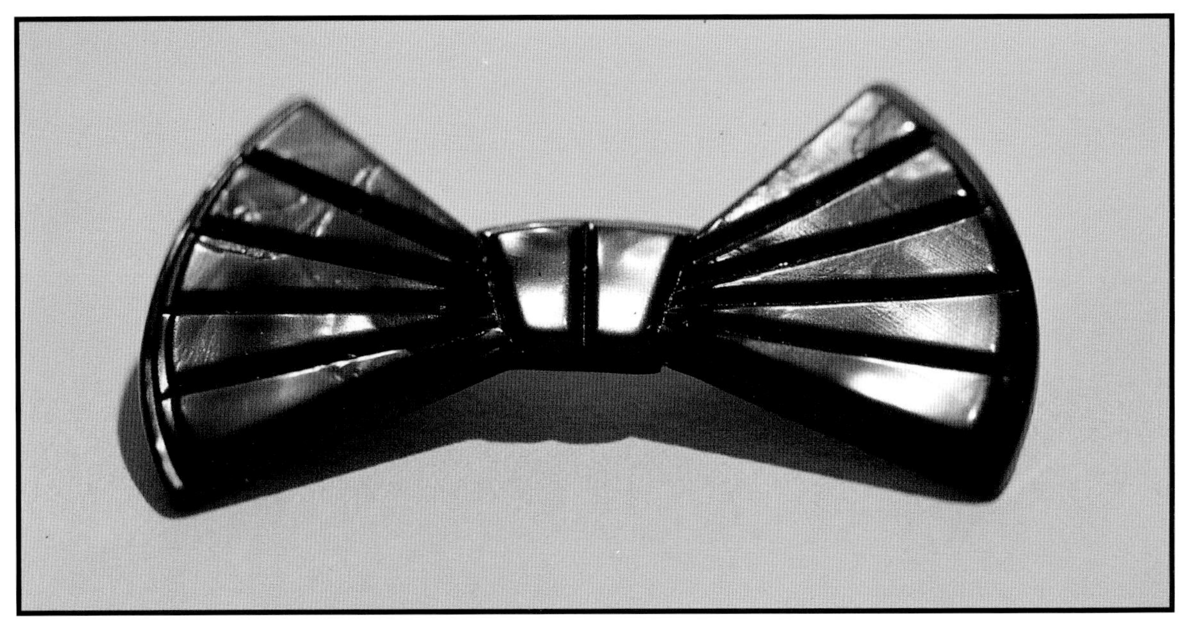

Silver glitter vest on a hanger pin, c. 1968-1980. 2". *Courtesy of private collector. Photo by Anita L. Bielecki.* $125-175.

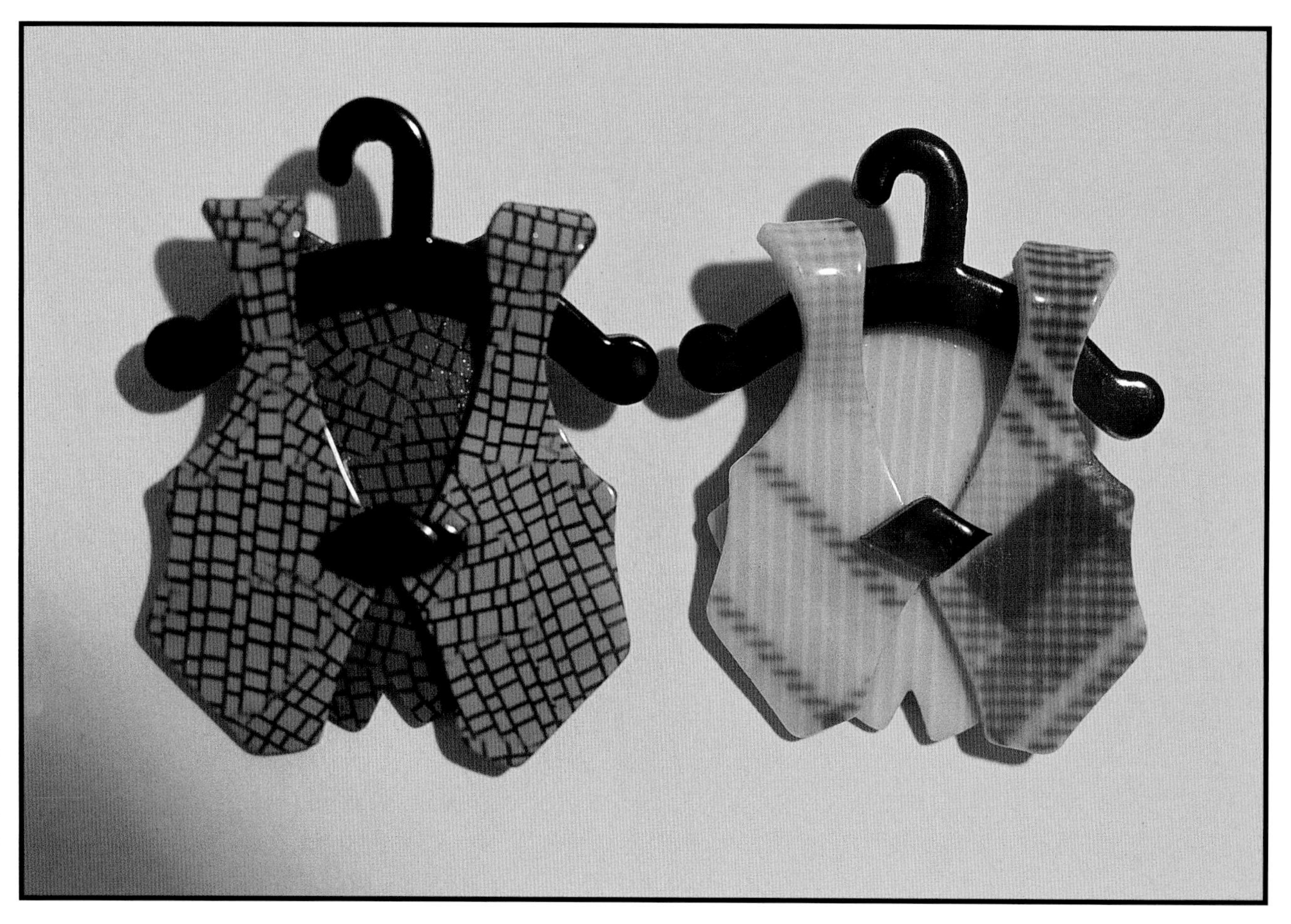

Pair of patterned vest on a hanger pins, c. 1968-1980. 2". *Courtesy of Remembrances of Things Past, Provincetown, MA. and author.* $125-175 each.

Pair of Rolls Royce car pins, c. 1968-1980. 2 1/4". *Courtesy of OhMy!!* $125-175 each.

Black Rolls Royce car pin with tortoise trim, c. 1968-1980. 2 1/4". *Courtesy of author.* $125-175.

Pair of Rolls Royce car pins, c. 1968-1980. 2 1/4". *Courtesy of Rememberances of Things Past, Provincetown, MA.* $125-175 each.

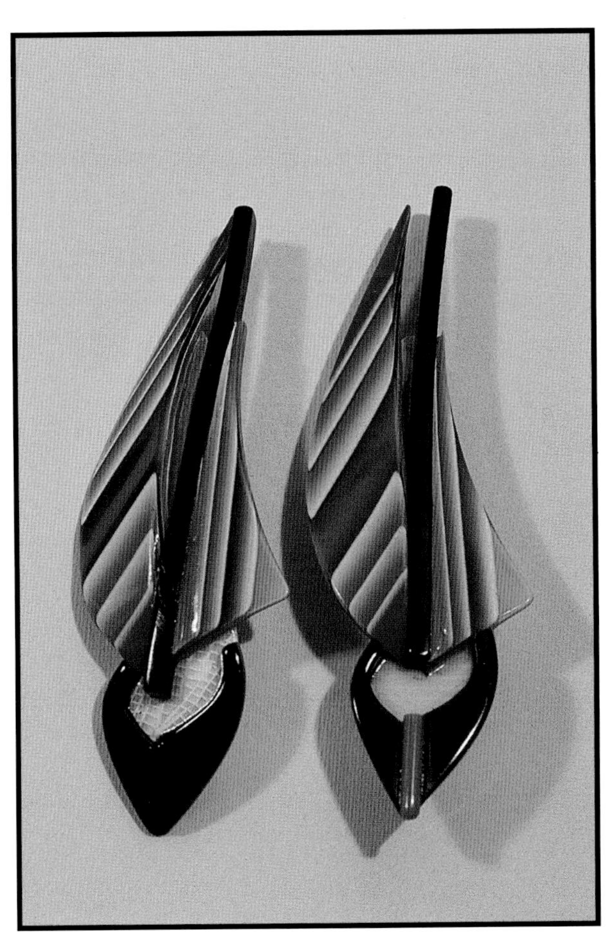

Pair of sailboat pins, c. 1968-1980. 3". *Courtesy of Dale Rhoads and author.* $125-175.

Tan and black umbrella pin, c. 1968-1980. 2 1/2". *Courtesy of Remembrances of Things Past, Provincetown, MA.* $125-175.

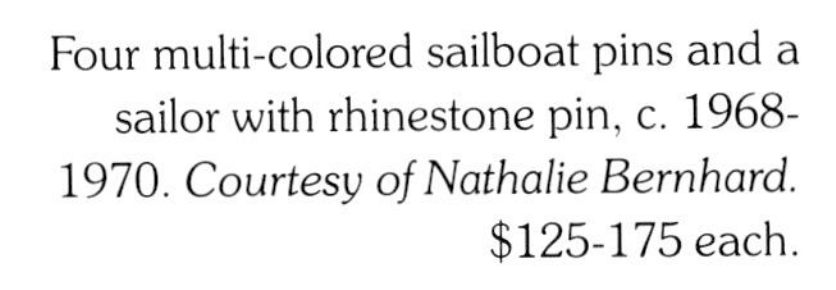

Four multi-colored sailboat pins and a sailor with rhinestone pin, c. 1968-1970. *Courtesy of Nathalie Bernhard.* $125-175 each.

Tiny pink patterned handbag with ivory closure pin, c. 1968-1980. 7/8". *Courtesy of Private Collector. Photo by Anita L. Bielecki.* $125-175.

Glitter umbrella pin, black and ivory tiny handbag pin, and envelope with feather pin, c. 1968-198 *Courtesy of OhMy!!* $125-175 each.

Red and black book pin, c. 1968-1980. 1 1/8". *Courtesy of OhMy!!* $125-175.

Open book pin, c. 1968-1980, *Courtesy of Nathalie Bernhard.* $125-175.

Small and large pins of crossed tennis rackets, c. 1968-1980. 2 3/4", 3 1/4". *Courtesy of Nathalie Bernhard.* $125-175 each.

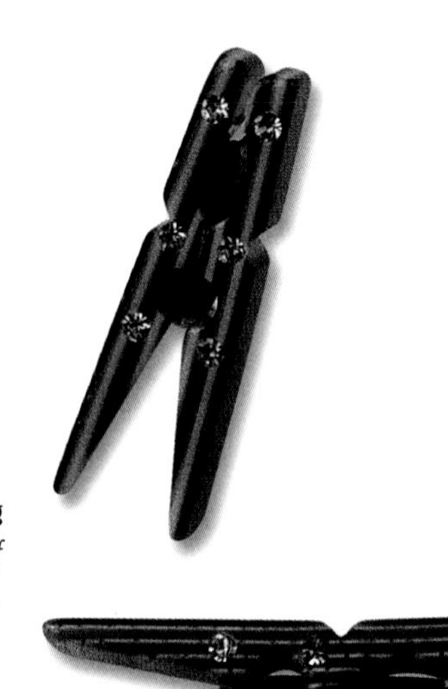

Pair of gray and white clothes peg pins, c. 1968-1980. *Courtesy of Nathalie Bernhard.* $125-175 each.

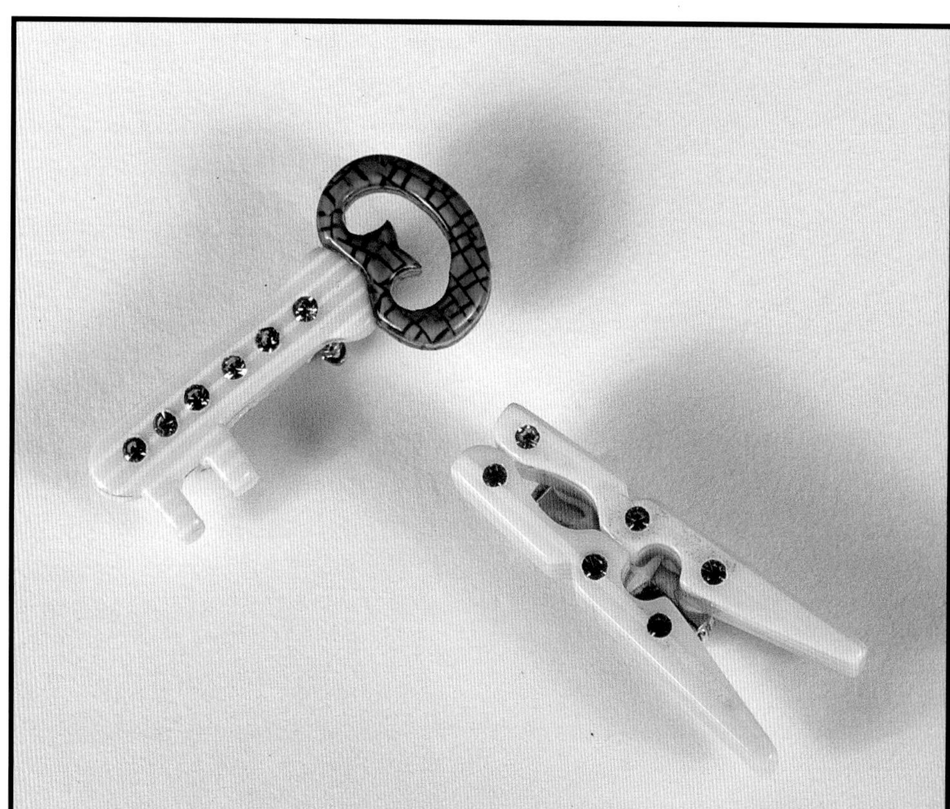

Ivory clothes peg pin and a key pin, both with rhinestones, c. 1968-1980. *Courtesy of Nathalie Bernhard.* $125-175 each.

Rhinestone hand-shaped pin, c. 1968-1980. 1 7/8". *Courtesy of OhMy!!* $125-175.

Tiny teapot pin, c. 1968-1980. 1". *Courtesy of Dale Rhoads.* $125-175.

Chapter 12

Bracelets and Rings

Pair of colored bangle bracelets with ivory lace, c. 1970s. *Courtesy of Gail Crockett.* $600-700 each.

Group of six multi-color layered stretch bracelets, c. 1970s, *courtesy of OhMy!!* $325-395 each.

Two layered stretch bracelets: one orange and ivory, the other brown, orange, and ivory, c. 1970s, *courtesy of OhMy!! and Remembrances of Things Past, Provincetown, MA.* $325-395 each.

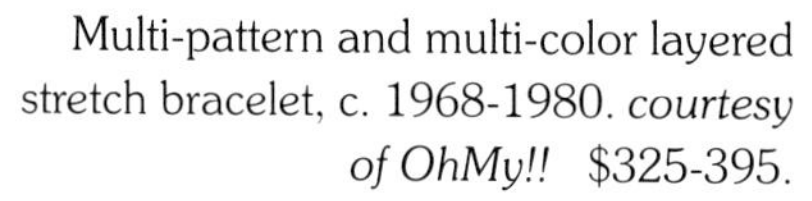

Multi-pattern and multi-color layered stretch bracelet, c. 1968-1980. *courtesy of OhMy!!* $325-395.

Layered multi-color stretch bracelet, c. 1968-1980. *Courtesy of OhMy!!* $325-395.

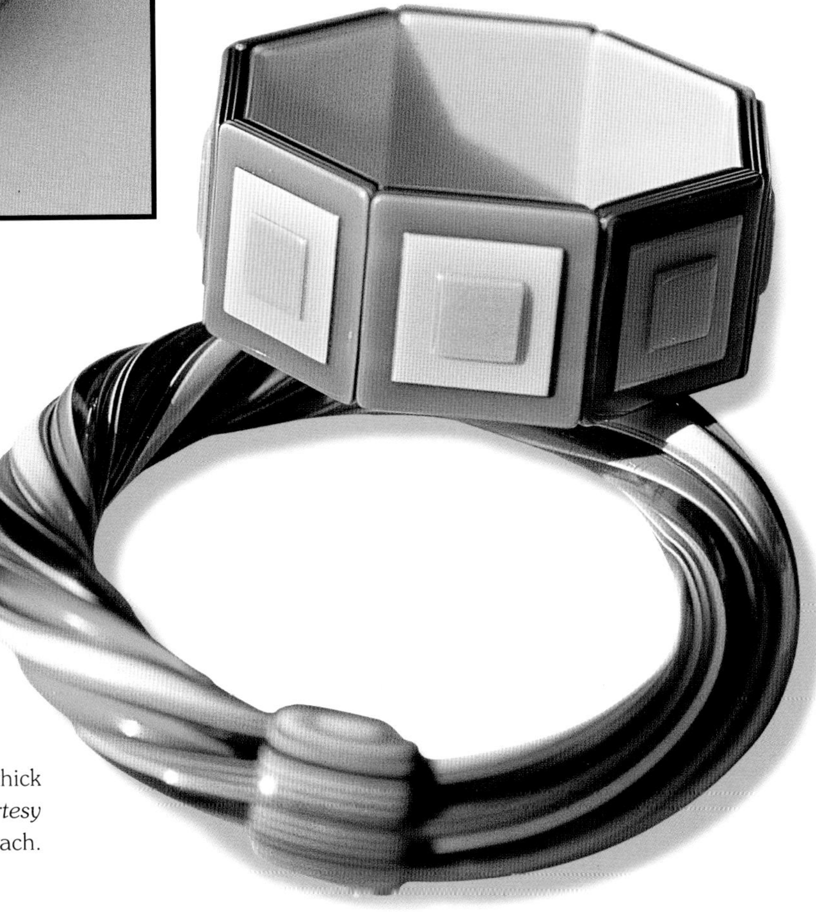

Layered green and yellow stretch bracelet and thick green and ivory twist bracelet, c. 1968-1980. *Courtesy of Nathalie Bernhard.* $175-375 each.

Pair of layered and carved multi-color bracelets, c. 1970s. *Courtesy of Dale Rhoads and author.* $600-700 each.

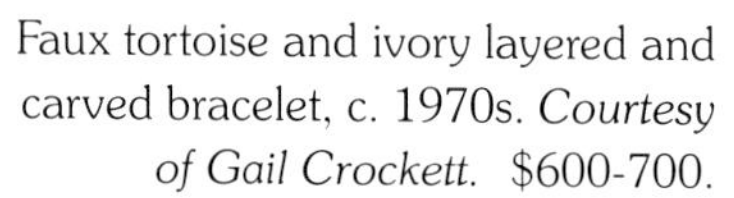

Faux tortoise and ivory layered and carved bracelet, c. 1970s. *Courtesy of Gail Crockett.* $600-700.

Layered and carved faux tortoise and ivory bracelet and exceptional faux tortoise layered snake bracelet, c. 1970s. *Courtesy of Dale Rhoads.* Carved bracelet $500-600. Snake Bracelet $150-195.

Brown textured cuff bracelet, c. 1968-1980. *Courtesy of Remembrances of Things Past, Provincetown, MA.* $175-275.

Bright red and green layered cuff bracelet, c.1970s. *Courtesy of Remembrances of Things Past, Provincetown, MA.* $200-275.

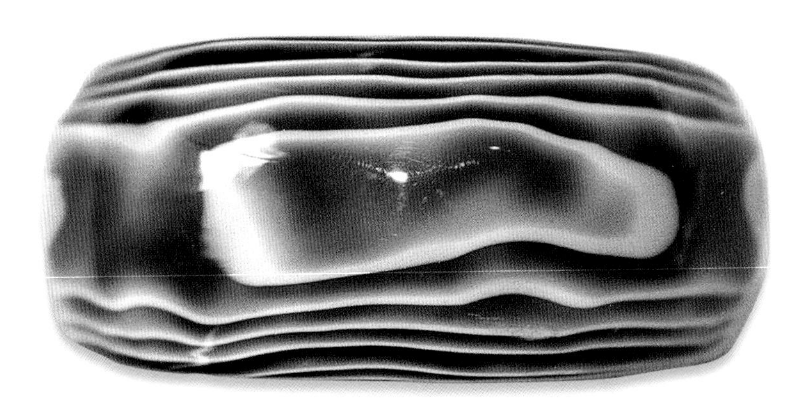

Rare faux tortoiseshell layered cuff bracelet, c. 1970s. *Courtesy of Nathalie Bernhard.* $200-275.

Pearlized striped cuff bracelet, c. 1968-1980. *Courtesy of Remembrances of Things Past, Provincetown, MA.* $175-275.

Textured cuff bracelets, pink and brown, c. 1968-1980. *Remembrances of Things Past Provincetown, MA.* $125-175 each.

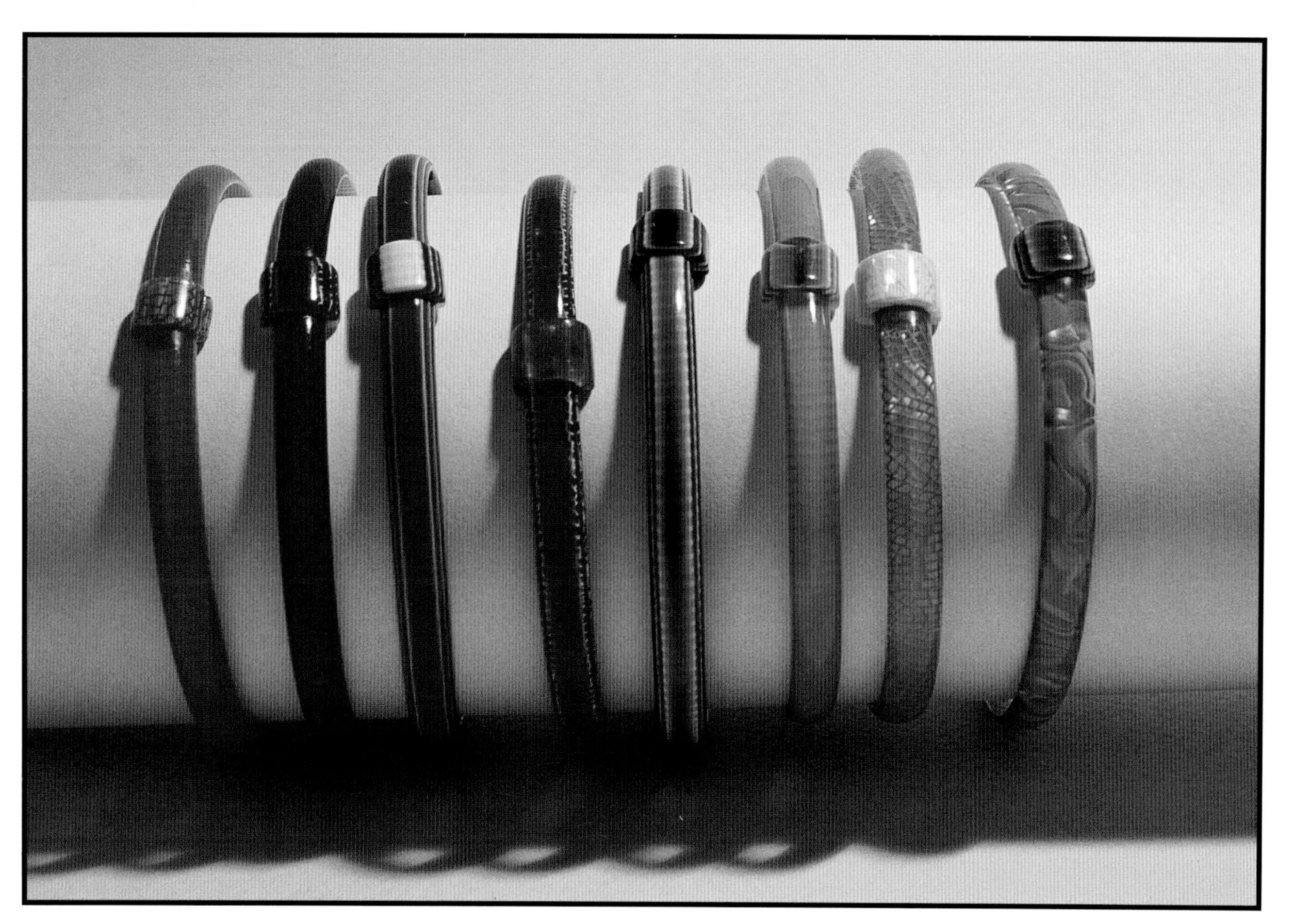

Collection of multi-color slim bangle bracelets, c. 1968-1980. *Remembrances of Things Past, Provincetown, MA.* $100-175 each.

Two bangle bracelets: double twist and slim striped, c. 1968-1980. *Remembrances of Things Past, Provincetown, MA.* $100-175 each.

Collection of slim bangle bracelets in ivory, silver, tortoise, and black, c. 1968-1980. *Remembrances of Things Past, Provincetown, MA.* $100-175 each.

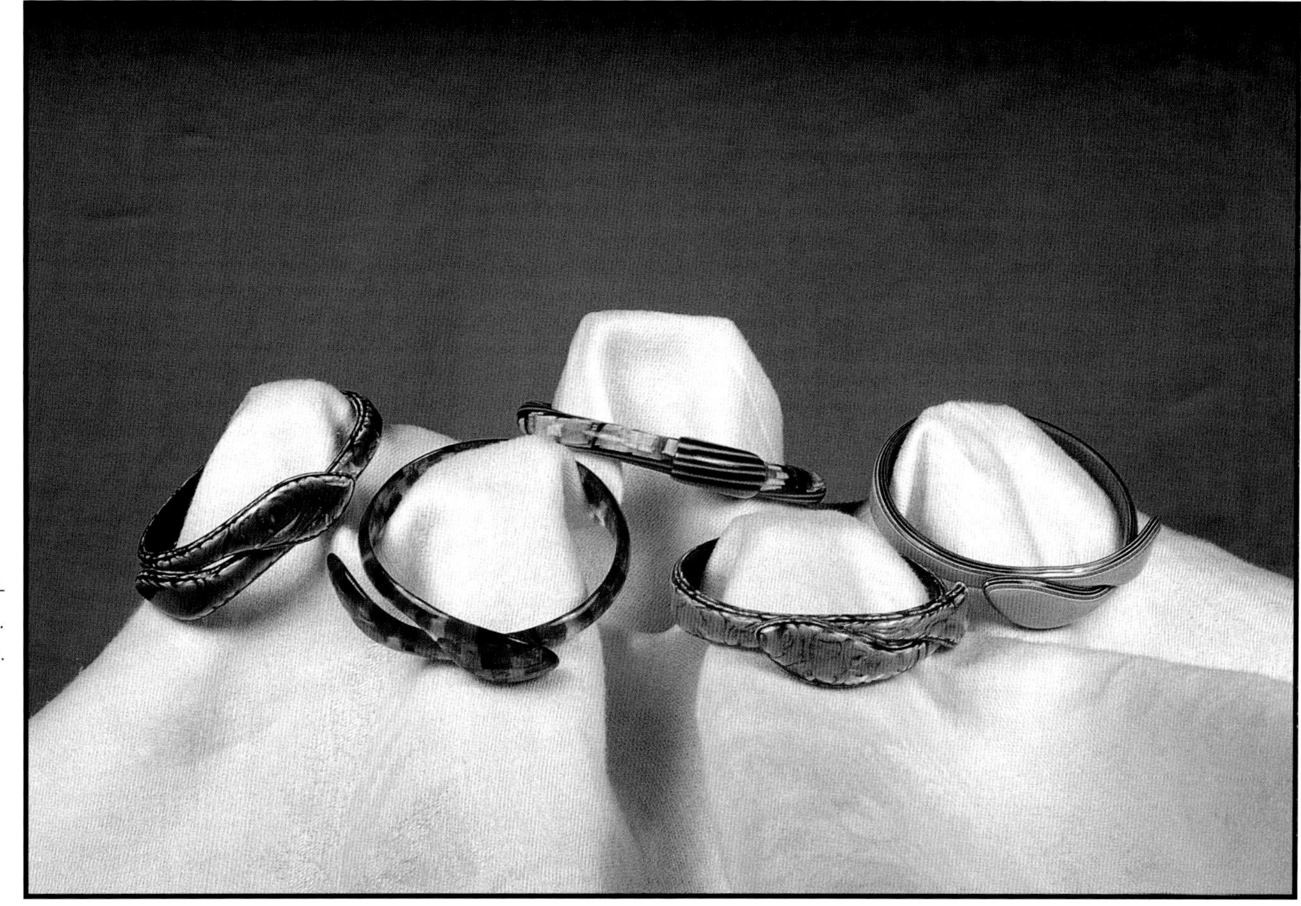

Five slim, multi-color, snake bangle bracelets, c. 1968-1980. *Courtesy of private collector. Photo by Anita L. Bielecki.* $100-175 each.

Six wide, multi-color, snake bangle bracelets, c. 1968-1980, *Courtesy of OhMy!! and author.* $100-175 each.

Five various slim circle bangle bracelets, c. 1980-1990s. *Courtesy of Dale Rhoads and author.* $100-175 each.

Four thick twist bracelets, ivory and black, c. 1968-1980. *Courtesy of OhMy!! and Remembrances of Things Past, Provincetown, MA.* $170-$275 each.

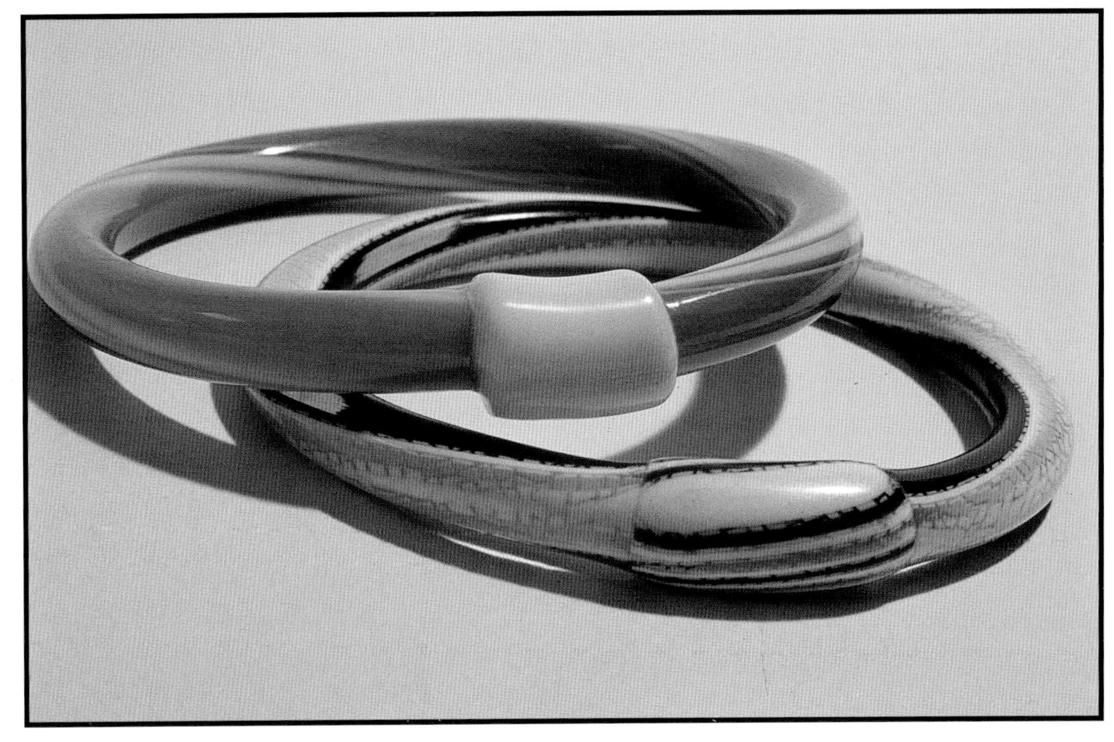

Two multi-color bangle bracelets, c. 1968-1980. *Courtesy of Remembrances of Things Past, Provincetown, MA.* $100-175 each.

Two striped multi-color bracelets, one twisted and one flat, c. 1968-1980. *Courtesy of Dale Rhoads and author.* $175-275 each.

Four multi-color square finger rings, c. 1968-1980. *Courtesy of Nathalie Bernhard.* $50-75 each.

Three multi-color striped, triangle rings, c. 1968-1980. *Courtesy of Remembrances of Things Past, Provincetown, MA.* $50-75 each.

Four multi-color striped, round band rings, c. 1968-1980. *Courtesy of Remembrances of Things Past, Provincetown, MA.* $50-75 each.

Collection of multi-color striped, round band rings, c. 1968-1980. *Courtesy of Nathalie Bernhard.* $50-75 each.

Chapter 13

Necklaces and Earrings

Large bead necklace of black and ivory with ivory spacers, c. 1968-1980. *Courtesy of Remembrances of Things Past, Provincetown, MA.* $225-400.

Large bead necklace of black and ivory with gold spacers, c. 1968-1980. *Courtesy of Dale Rhoads.* $225-400.

Bead necklace of black and hot pink flat discs with geometric pendent, c. 1968-1980. *Courtesy of Nathalie Bernhard.* $300-400.

Necklace of black round, and shaped and layered beads with silver spacers, c. 1968-1980. *Courtesy of Nathalie Bernhard.* $300-400.

Multi-color bead necklace with gold spacers, c. 1968-1980. *Courtesy of Dale Rhoads.* $300-400.

Rare and large red Aztec style pendent on black cord, c. 1968-1980. 5 1/2". *Courtesy of Gail Crockett.* $500-700.

Rare necklace of a chain and two doves pendant, c. 1968-1980. *Courtesy of Nathalie Bernhard.* $225-400.

Flat disc necklace with matching earrings in aqua and black, c. 1968-1980. *Courtesy of Nathalie Bernhard.* $300-400 set.

Two flat disc necklaces in black and different multi-colors, c. 1968-1980. *Courtesy of OhMy!!* $225-300 each.

Black and white flat disc necklace, c. 1968-1980. *Courtesy of private collector. Photo by Anita L. Bielecki.* $225-300.

Patterned multi-color earrings in stacked block shape, c. 1968-1980. *Courtesy of OhMy!!* $100-150.

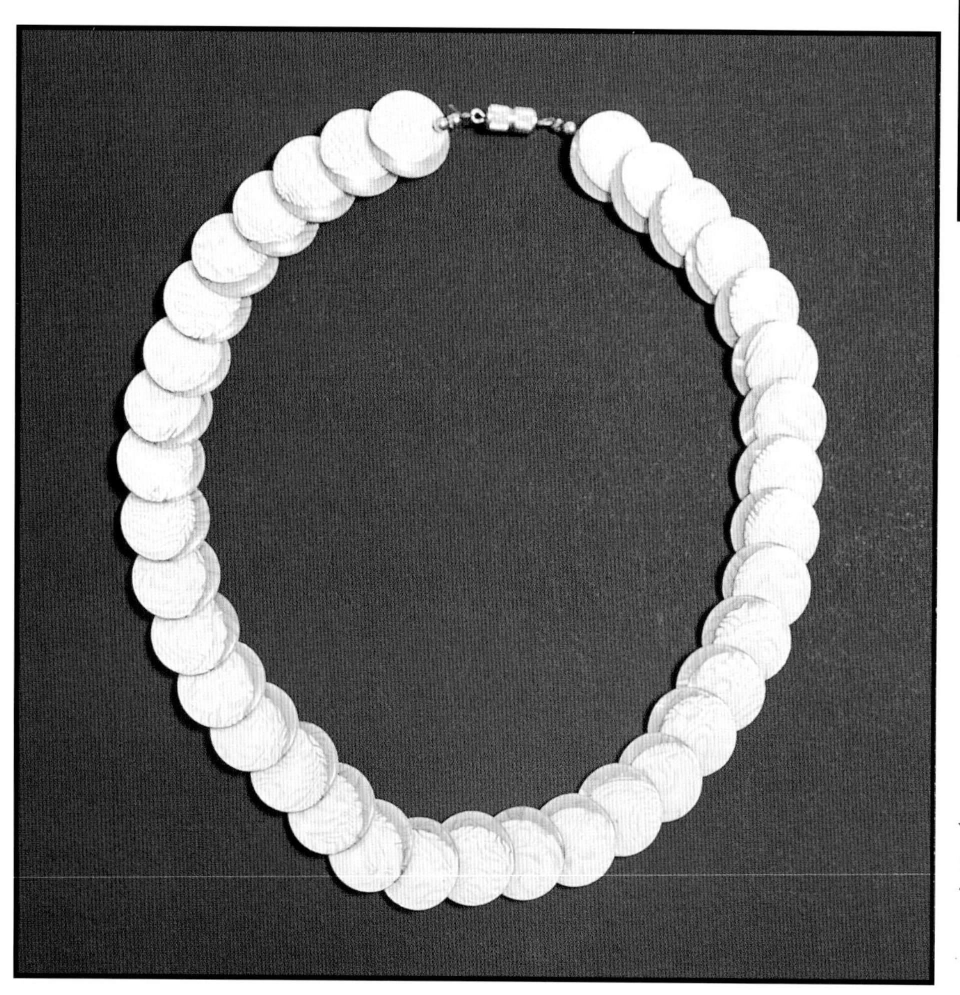

Yellow and ivory flat disc necklace, c. 1968-1980. *Courtesy of Remembrances of Things Past, Provincetown, MA.* $225-300.

Solid multi-color earrings in stacked block shape, c. 1968-1980. *Courtesy of OhMy!!* $100-150.

Green carved pinwheel ear clips, c. 1968-1980. *Courtesy of private collector. Photo by Anita L. Bielecki.* $75-150.

Red carved pinwheel ear clips, c. 1968-1980. *Courtesy of Nathalie Bernhard.* $75-150.

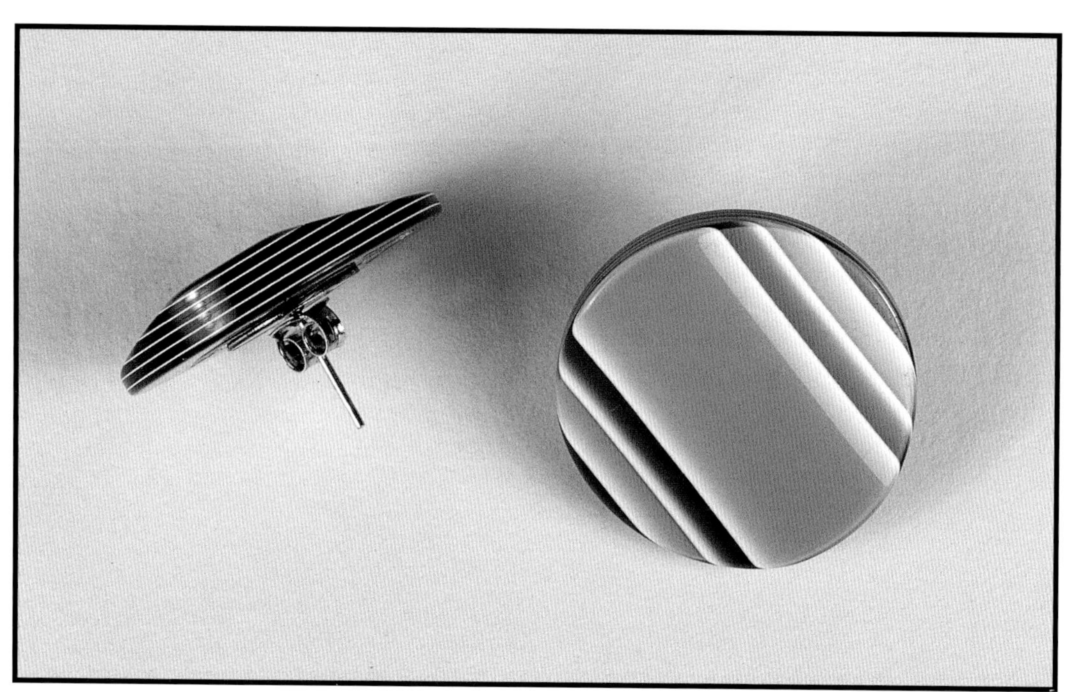

Round pierced earrings with layers cut through, c. 1968-1980. *Courtesy of private collector. Photo by Anita L. Bielecki.* $75-125.

Ear clips and matching ring with layers cut through, c. 1968-1980. *Courtesy of Nathalie Bernhard.* $150-300 set.

Three pair of red layered earrings, c. 1968-1980. *Courtesy of Dale Rhoads.* $100-150 each pair.

Three pair of earrings: pearlized multi-color squares, engraved hearts, and art deco style circles, c. 1968-1980. *Courtesy of Dale Rhoads.* $100-150 each pair.

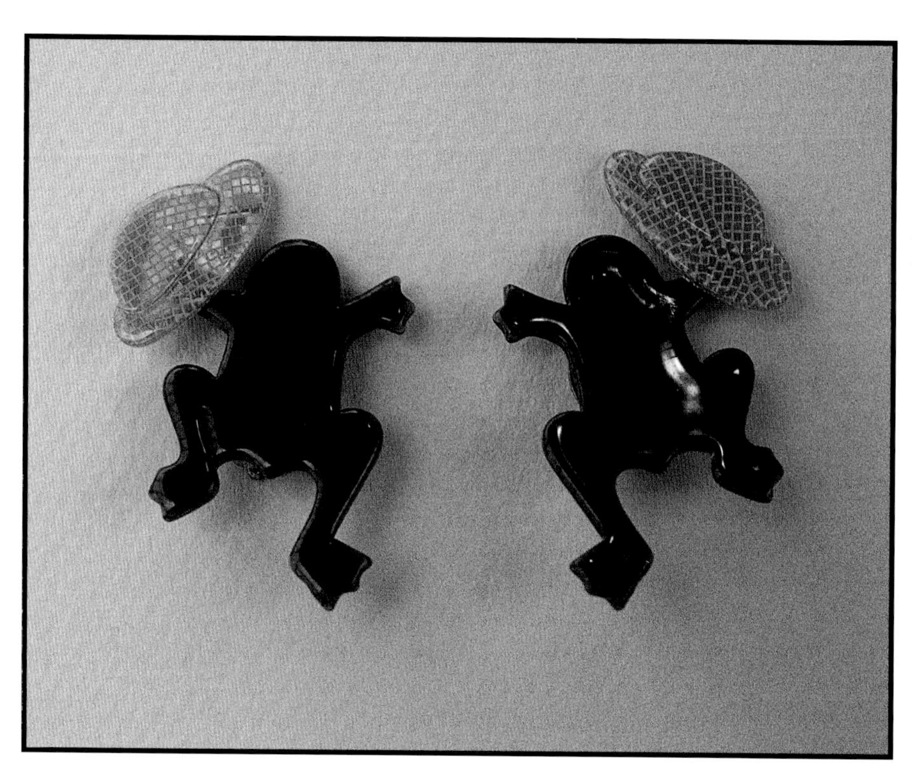

Frog with a pink hat earrings. c. 1990s.
Courtesy of Nathalie Bernhard. $100-150.

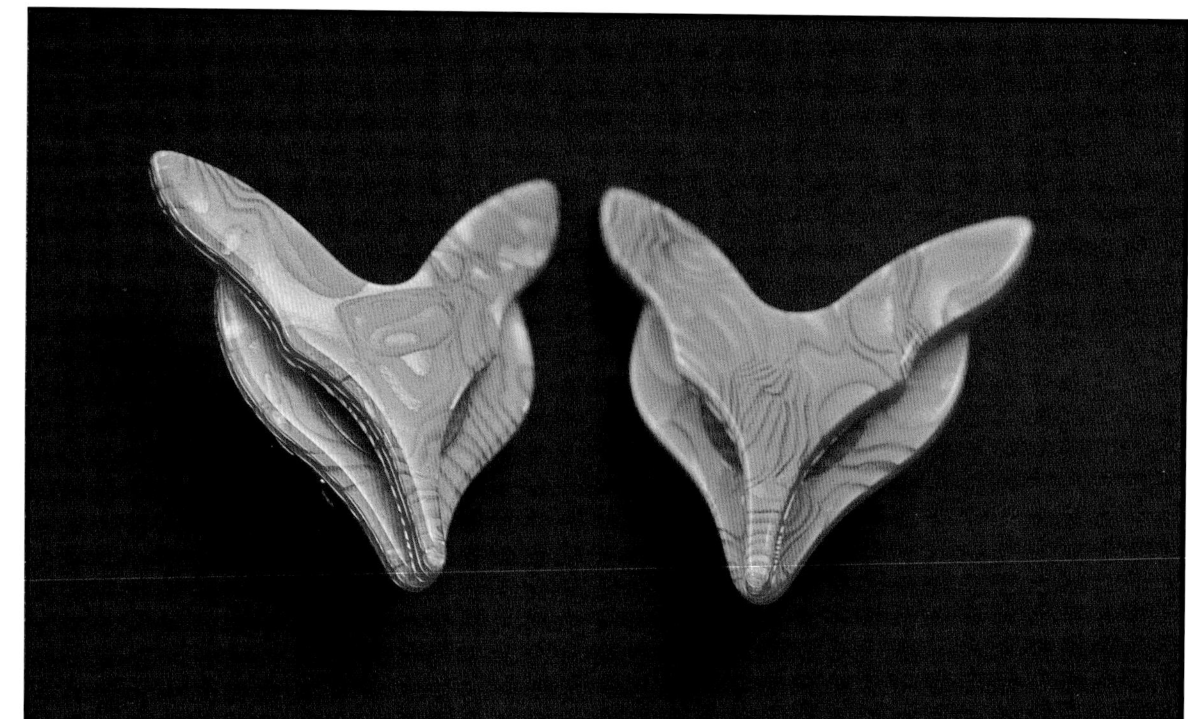

Pink fox head earrings, c. 1990s.
Courtesy of OhMy!! $100-150.

Ladybug earrings, c. 1990s. *Courtesy of Nathalie Bernhard.* $100-150.

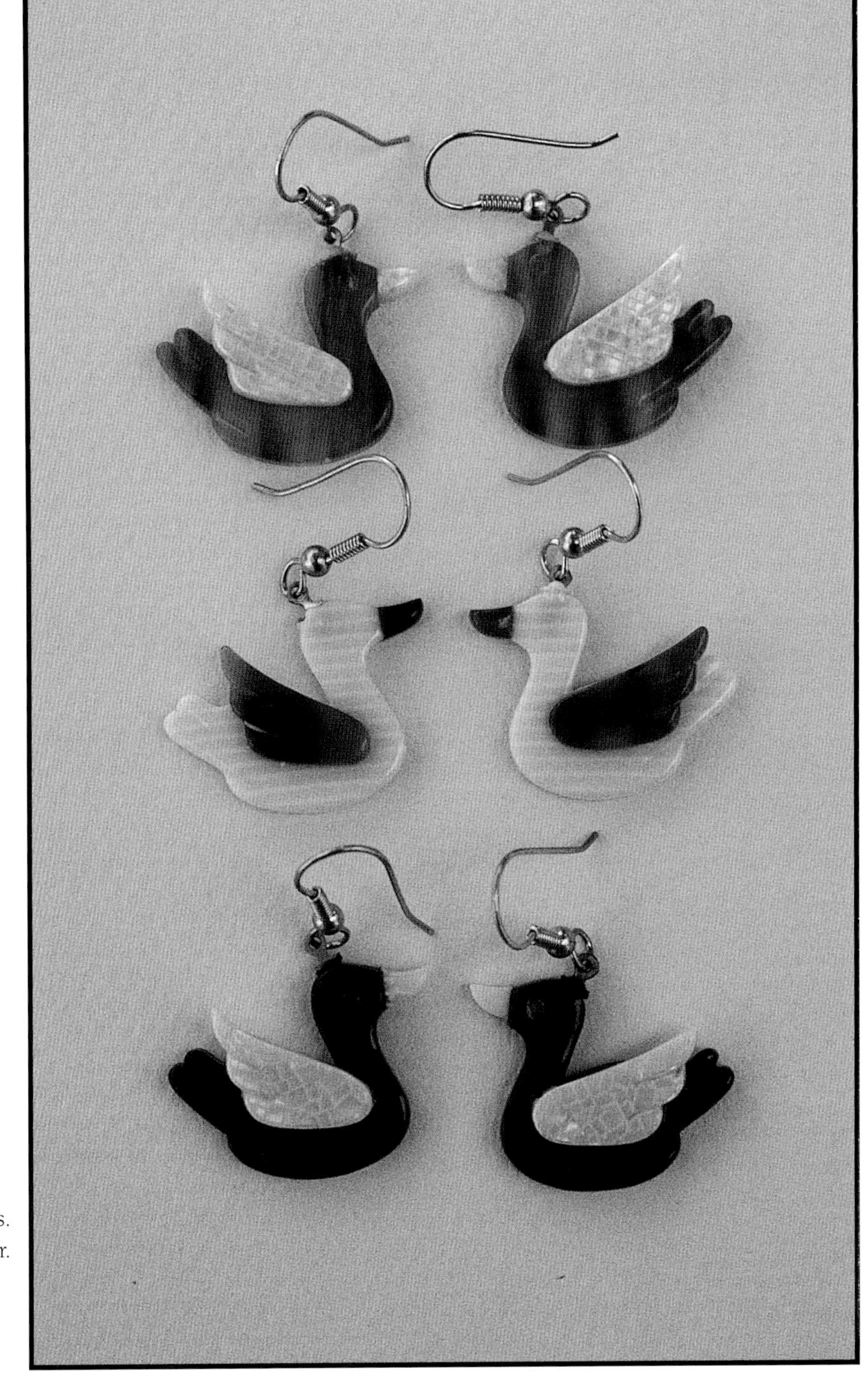

Three different pair of duck earrings, c. 1990s. *Courtesy of Nathalie Bernhard.* $75-125 each pair.

Three different multi-color cat face earrings, c. 1990s. *Courtesy of OhMy!! and Dale Rhoads.* $100-150 each pair.

Cat face earrings and matching pin, c. 1990s. *Courtesy of Dale Rhoads.* Earrings $100-150, Pin $125-175.

Chapter 14

Ornamental Hair Combs

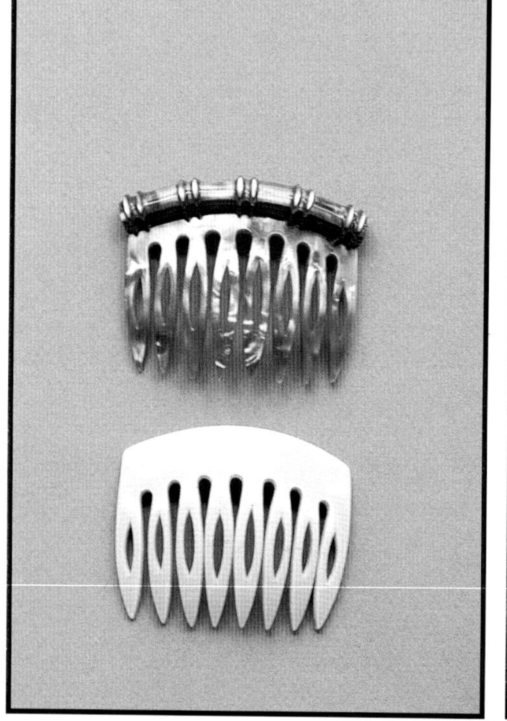

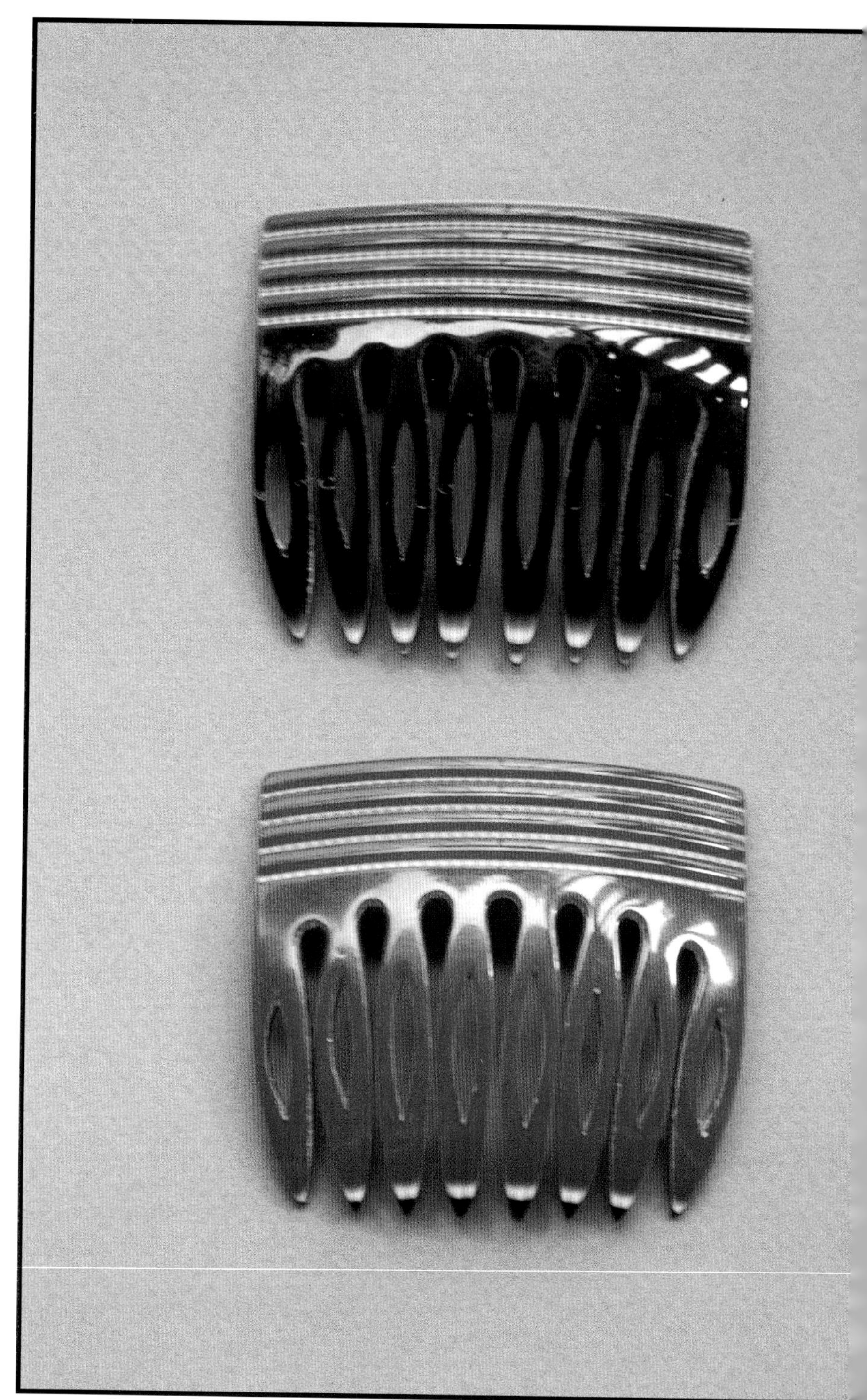

Pairs of hair combs, including multi-color floral, butterfly, flapper, striped, leaf, Egyptian eye, and bow design in carved and layered construction, c. 1968-1980. *Courtesy of Lea Stein Collection, Paris.*

Green hair comb, c. 1968-1980. *Courtesy of Nathalie Bernhard.* $125-175

Floral, multi-stripe, and lady bug hair combs. c. 1980-1990s. *Courtesy of Nathalie Bernhard.* $125-175 each.

Chapter 15

Accessories

Many of the compacts, perfume holders, and other accessories were designed exclusively for French cosmetic companies. Lea Stein also designed accessories such as jewel boxes, purse mirrors, belt buckles, and more for everyone.

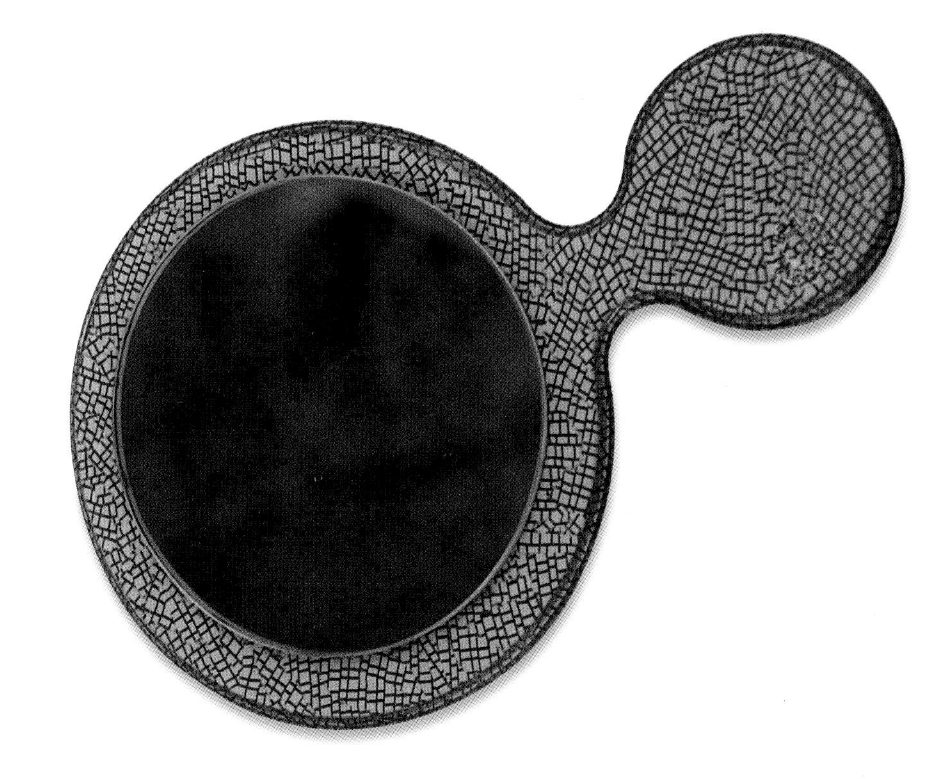

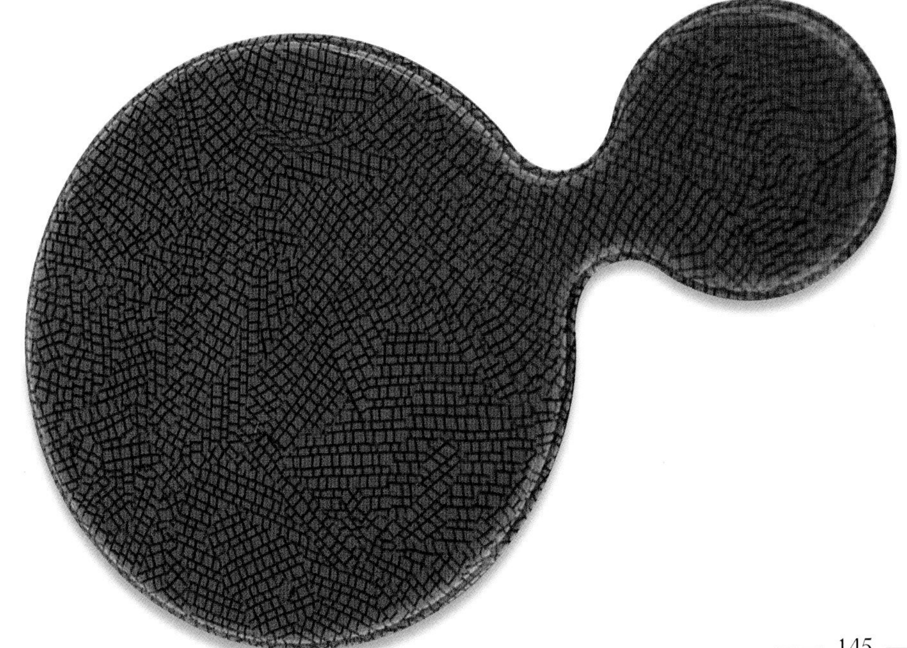

Purple hand mirror, front and back, c. 1970s. 5 1/2". *Courtesy of Nathalie Bernhard.* $125-225.

Purple layered picture frame, c. 1970s. 5 3/4". *Courtesy of Gail Crockett.* $200-400.

Purple jewel box with layered top in colored layers, c. 1970s. 6 1/2". *Courtesy of Gail Crockett.* $350-700.

Two views of a purple pivoting mirror, c. 1970s. Approx. 5 1/2" *Courtesy of Gail Crockett.* $200-400.

Pair of serigraphy belt buckles with cat face and flowers design, c. 1960s. *Courtesy of Nathalie Bernhard.* $200-300 each.

Faux tortoise decorated clasp. c. 1960s. *Courtesy of Nathalie Bernhard.* $175-275.

Three round serigraphy purse mirrors, c. 1975. Designed for French fashion designer Gudule. *Courtesy of Nathalie Bernhard.* $300-400 each.

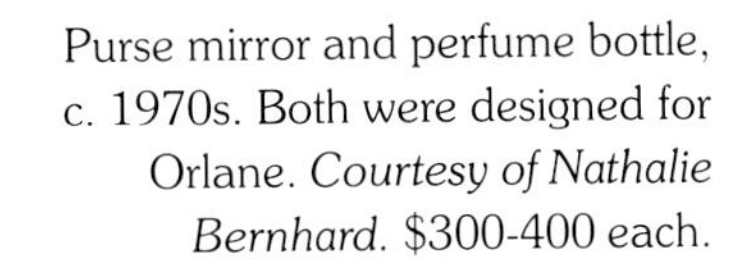
Purse mirror and perfume bottle, c. 1970s. Both were designed for Orlane. *Courtesy of Nathalie Bernhard.* $300-400 each.

Three cigarette lighter holders, c. 1975. These were designed to hold the disposable cigarette lighters sold in tobacco shops and were meant to be worn around the neck with the black cord. 2 1/2". *Courtesy of Nathalie Bernhard*. $100-200.

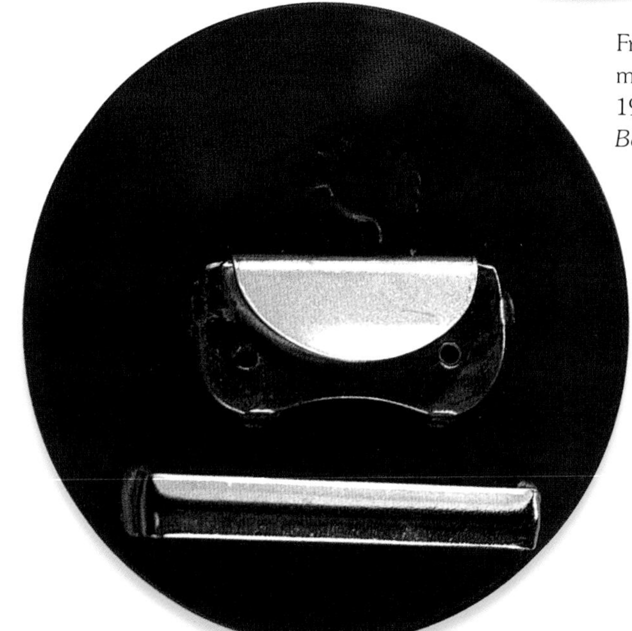

Front and back views of a multi-colored belt buckle, c. 1970s. *Courtesy of Nathalie Bernard*. $125-175.

Chapter 16

Geometric Designs

Group of geometric pins, c. 1968-1980.
Courtesy of OhMy!! $125-175 each.

Stacked cube pin in multi-colors, c. 1968-1980.
1 3/4" *Courtesy of OhMy!!* $100-150.

Multi-colored geometric cube pins, c. 1968-1980. 2 3/4" and 1 3/4". *Courtesy of OhMy!!* $125-175 each.

Black and white geometric cube pin, c. 1968-1980. 2 3/4". *Courtesy of OhMy!!* $125-175.

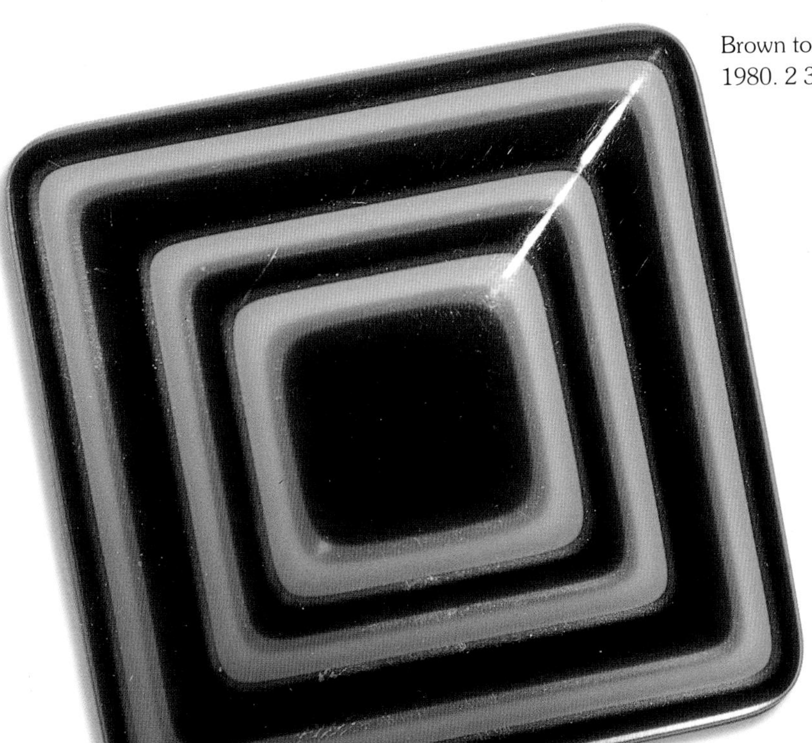

Brown tortoise and white geometric cube pin, c. 1968-1980. 2 3/4". *Courtesy of OhMy!!* $125-175.

Red geometric square pin, c. 1968-1980. 1 3/8" *Courtesy of OhMy!!* $125-175.

Tortoise and ivory geometric square pin, c. 1969-1980. 1 3/8". *Courtesy of OhMy!!* $125-175.

Triangle geometric pin, c. 1968-1980. 1 5/8". *Courtesy of OhMy!!* $125-175.

Triangle tortoise geometric pin, c. 1968-1980. 1 5/8" *Courtesy of OhMy!!* $125-175.

Multi-colored Emporium building geometric pin, c. 1968-1980. 2" *Courtesy of OhMy!!* $125-175.

Multi-colored chevron pin, c. 1968-1980. 1 3/4". *Courtesy of Author.* $100-150.

Pair of multi-colored chevron pins, c. 1968-1980. 1 3/4" *Courtesy of OhMy!!* $100-150 each.

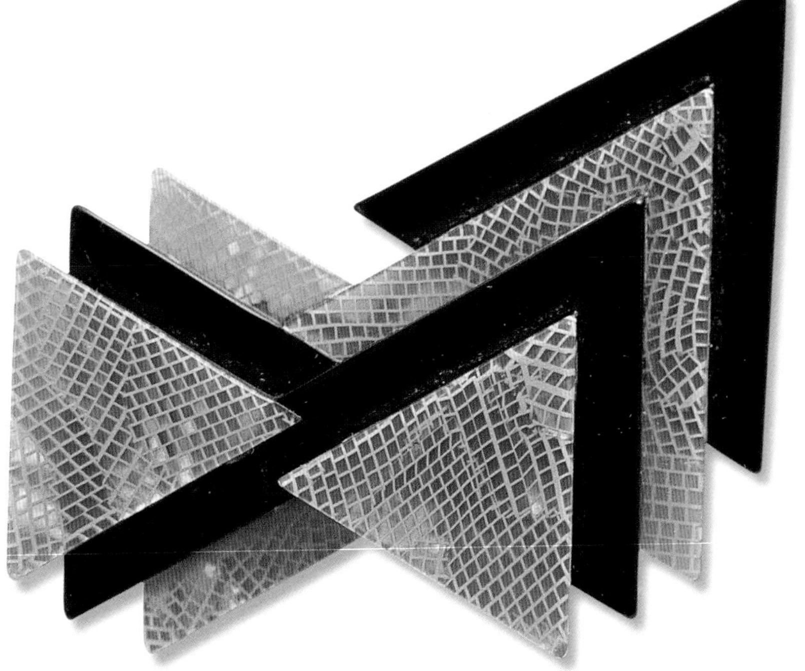

Black and gold stacked triangles pin, c. 1968-1980. 2 1/2". *Courtesy of Dale Rhoads.* $125-175.

Three tortoise and ivory geometric pins, c. 1968-1980. 2 1/2". *Courtesy of Remembrances of Things Past, Provincetown, MA.* $125-175 each.

Two diamond shaped geometric pins, c. 1968-1989. 1 3/8". *Courtesy of OhMy!!* $100-150 each.

Ivory diamond geometric pin, c. 1968-1980. 1 3/8" *Courtesy of Remembrances of Things Past, Provincetown, MA.* $125-175.

Engraved diamond bowl geometric pin, c. 1968-1980. 1 3/8". *Courtesy of OhMy!!* $125-175.

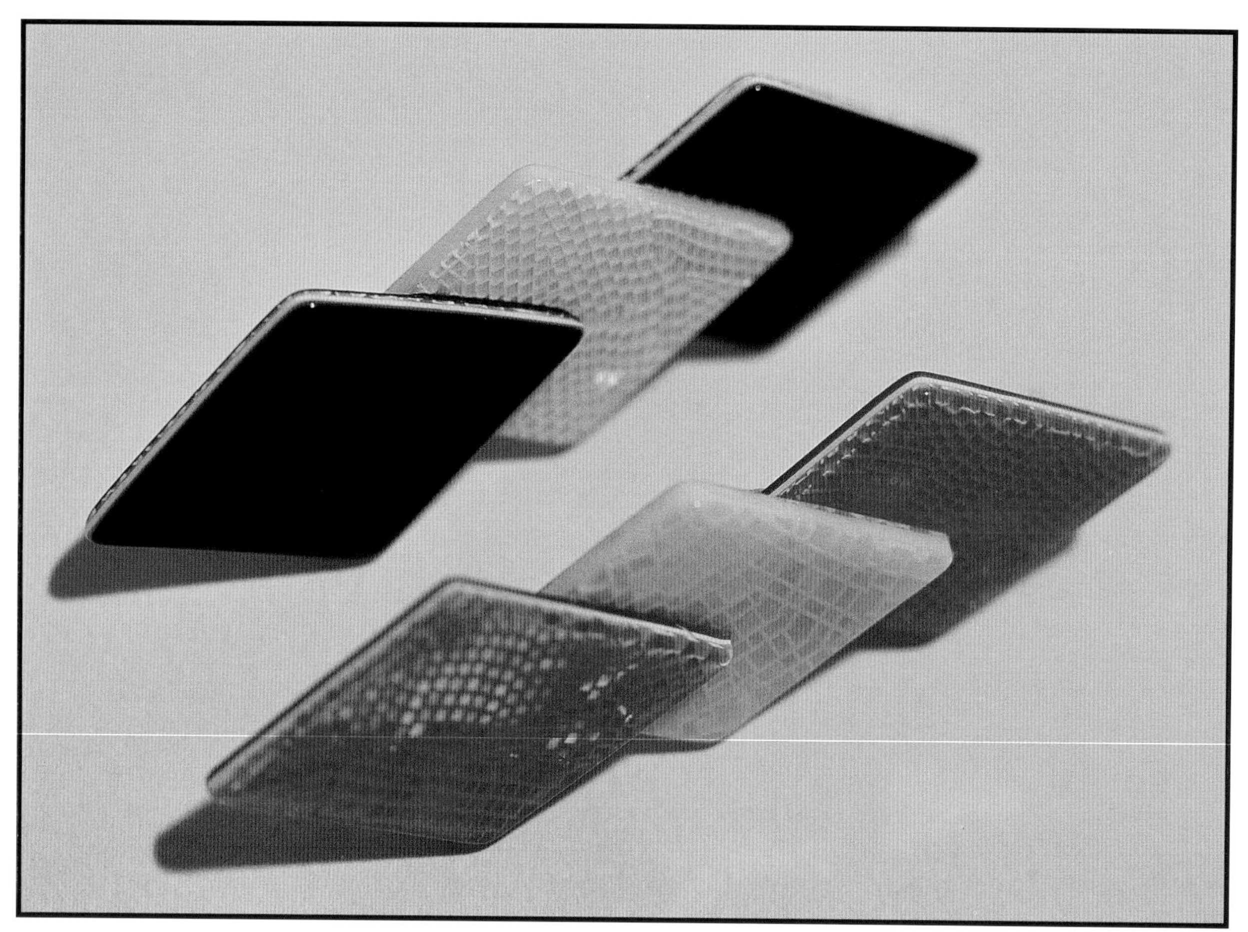

Two pins. each of three multi-color diamonds, c. 1968-1980. 3 1/4". *Courtesy of OhMy!!* $125-175.

Multi -color geometric pin, c. 1968-1980. 2 1/2". *Courtesy of Dale Rhoads.* $125-175.

Fan-shaped, multi-color pin, c. 1968-1980. 1 7/8". *Courtesy of OhMy!!* $125-175.

Half moon pin, c. 1968-1980. *Courtesy of Remembrances of Things Past, Provincetown, MA.* $100-150.

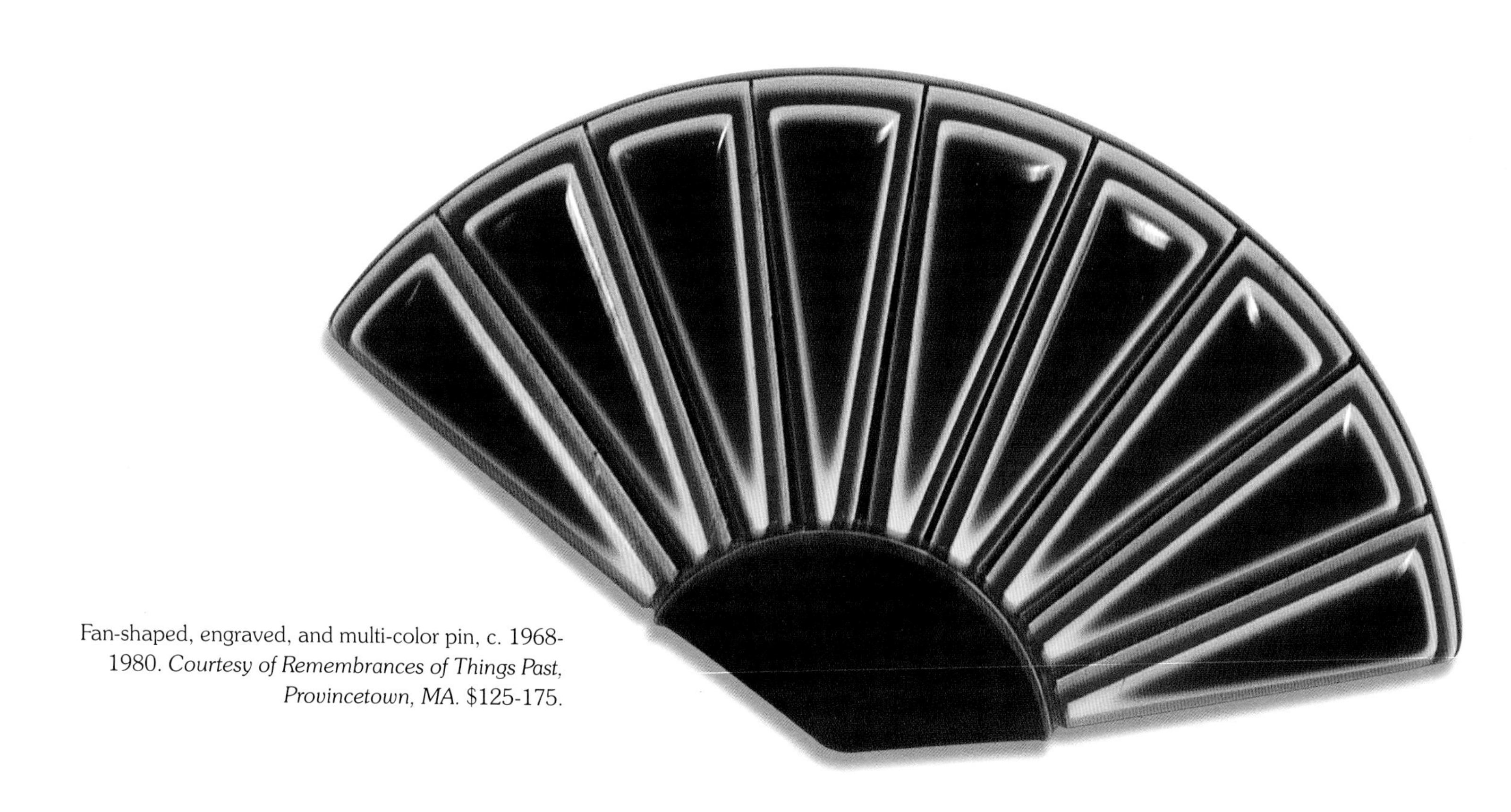

Fan-shaped, engraved, and multi-color pin, c. 1968-1980. *Courtesy of Remembrances of Things Past, Provincetown, MA.* $125-175.